IntechOpen

Supporting open minds since 2005

Vascular Biology - Selection of Mechanisms and Clinical Applications
http://dx.doi.org/10.5772/intechopen.78429
Edited by Marcelo González

Contributors
Xhevdet Krasniqi, Hajdin Citaku, Sabyasachi Sen, Eric Nylen, Shruti Gandhi, Dai-Min Zhang, Xiao-Xin Jiang, Yan-Rong Zhu, Peng Ye, Shao-Liang Chen, Chieh-Hsi Wu, Chun-Hsu Pan, Ming-Jyh Sheu, Weilue He, Maria Kwesiga, Eyerusalem Gebreyesus, Sijia Liu, Vera Bernardino, Ana Rodrigues, Ana Lladó, Melissa Fernandes, António Panarra, Ignacio Vila, Beatriz Brea-García, Iván Couto Gonzalez, Abhishek Patil, Isha Sood, Mariane Manso, Thaisa Macarini De Faria, Ana Paula Urzedo, Muhammad Shariq Mukarram, Muhammad Ishaq Ghauri, Rosa Pereira, Marilia Ambiel Dagostin, Eda Öksüm Solak, Murat Borlu, Marcelo González, José Carlos Rivas

First published in London, United Kingdom, 2020 by IntechOpen
IntechOpen is the global imprint of INTECHOPEN LIMITED, registered in England and Wales, registration number: 11086078, 5 Princes Gate Court, London, SW7 2QJ, United Kingdom
Printed in Croatia

British Library Cataloguing-in-Publication Data
A catalogue record for this book is available from the British Library

Additional hard and PDF copies can be obtained from orders@intechopen.com

Vascular Biology - Selection of Mechanisms and Clinical Applications
Edited by Marcelo González
p. cm.
Print ISBN 978-1-83969-035-8
Online ISBN 978-1-83969-036-5
eBook (PDF) ISBN 978-1-83969-037-2

For EU product safety concerns:
IN TECH d.o.o., Prolaz Marije Krucifikse Kozulić 3, 51000 Rijeka, Croatia, info@intechopen.com or visit our website at intechopen.com.

Meet the editor

Dr. Marcelo González studied Biochemistry at the Universidad de Concepción, Chile, and has a Doctoral degree in Physiological Sciences from the Universidad Católica de Chile. From 2009 he has been working at the Universidad de Concepción, first in the Department of Physiology and now in the Department of Obstetrics and Gynecology. His research interest include the description of placental vascular regulation mechanisms, both in physiological conditions and in pregnancy diseases such as gestational diabetes and preeclampsia. In his laboratory, scientists use different methods of cellular biology, molecular biology, biochemistry, and vascular reactivity to elucidate the mechanisms of vascular regulation from the molecular interactions to organ function. Recently, his work has focused on the study of potassium channel regulation and its interaction with nitric oxide signaling in the human placenta, to determine the potential alterations of these mechanisms in gestational diabetes. He has published 32 articles in WOS journals and 4 book chapters.

Contents

Preface

According to the World Health Organization (WHO), the two main causes of death in the world are directly related to cardiovascular system disorders, ischemic heart disease, and stroke. Clinical evidence demonstrates the relevance of knowledge about vascular biology, from molecular mechanisms to clinical applications.

This book includes chapters that describe mechanisms of vascular regulation, pathophysiological evidence of vascular diseases, and some clinical application microvascular surgery. Related to molecular mechanisms of regulation of vascular function, the book is focused on nitric oxide signaling and the activity of potassium channels. Both mechanisms are related to the function of the endothelial cells, a key cell-type in the homeostasis of the vascular system. Also reviewed is the effect of physical activity and insulin on endothelial function, as an essential physiological factor to maintain the vascular health.

In relation to vascular diseases, the second section of the book covers the disease of coronary arteries, mainly related with structural alteration, and also the alterations of the vascular system in systemic sclerosis, an autoimmune disease that affects several organs. This section includes two chapters related to vasculitis, which is the inflammation of blood vessels. The first chapter focuses on the pathophysiology and clinical characteristics and the second chapters describes the pharmacological treatments.

The third section of the book is focused on the microvasculature, mainly centered in the nail fold capillaroscopy, a simple, low-cost method that is extremely important in the evaluation of patients with rheumatic spectrum diseases. Finally, the last chapter presents a review of critical steps and recommendations for the transplant of microvascular tissues for improvement of functional and aesthetic results in microvascular surgery.

This book is an international effort of collaboration, with the purpose to create an academic tool for students of medical sciences or people interested in learning about vascular biology. I invite the readers to enjoy the chapters and contact the authors in the case of needing more information.

Marcelo González
Laboratorio de Investigación Materno-Fetal (LIMaF),
Departamento de Obstetricia y Ginecología,
Facultad de Medicina,
Universidad de Concepción,
Chile

Section 1

Physiological and Pathophysiological Mechanisms

Chapter 1

Nitric Oxide and Oxidative Stress-Mediated Cardiovascular Functionality: From Molecular Mechanism to Cardiovascular Disease

Weilue He, Maria Paula Kwesiga, Eyerusalem Gebreyesus and Sijia Liu

Abstract

The underlying pathology of most cardiovascular diseases (CVDs) such as coronary artery disease, high blood pressure, and stroke involves decreased cardiovascular contractility and anatomic alterations in cardiovascular structures. Nitric oxide (NO) regulates vascular tone and contractile function of myocardium and maintains blood vessel homeostasis. Interestingly, the effect of NO is like a double-edged sword in the body. Insufficient NO causes hypertension and atherosclerosis, while an overproduction of NO may foster inflammation and cause heart infarction and shock. In addition, growing evidences have shown that oxidative stress plays pivotal roles in the initiation and progression of CVDs. This chapter will discuss in detail the roles NO plays in the cardiovascular system under both physiological and pathological conditions. We will focus on: (1) the molecular mechanism of cardiovascular contraction, (2) NO/Ca^{2+}-induced muscle relaxation, (3) NO-related structural change in blood vessels, and (4) redox balance in the cardiovascular system. The relationships between these molecular mechanisms and the characteristics of CVDs will be highlighted.

Keywords: cardiovascular diseases, muscle contraction and relaxation, cytoskeleton, nitric oxide, nitroso-redox balance

1. Introduction

Cardiovascular diseases (CVDs), i.e., ischemic heart disease and stroke, remain the leading cause of death in the past decades around the world, especially in the developed countries [1]. CVDs can start from risk factors that may cause local vascular lesion and end up with systematic complications, which lead to organ failure and death. Thus, understanding the biochemistry of events involved in the whole process of CVD progression is crucial to prevent and treat the disease.

Epidemiological data show that various factors are associated with the increase of cardiovascular morbidity and mortality, including hypertension, smoking, hypercholesterolemia, diabetes mellitus, obesity, stress, low fruit and vegetable dietary, lack of regular exercise, and abnormal sleep [2]. Current therapeutic strategies mainly focus on reducing patients' blood pressure, restoring redox balance, controlling cholesterol, and implementing physical activity programs [3]. In this chapter, we explore the physiological and pathological events in the cardiovascular system from the molecular biology's perspectives. Molecular mechanism of muscle contraction and relaxation in the cardiovascular system will be discussed first. Then, we will delve into the biological effects of Nobel Prize molecule nitric oxide (NO), the most important vasodilator in the body. In addition, due to the inspiring clinical outcomes of using isosorbide dinitrate (an NO stimulus) and hydralazine (an antioxidant) to treat patients with symptomatic congestive heart failure [4], we will also discuss how nitroso-redox balance mediates cardiovascular functions.

2. Muscle contraction and relaxation

2.1 Sliding filament theory

Skeletal, cardiac, and smooth muscles have different structures and regulatory mechanisms, but they share the same molecular mechanism of contraction and relaxation, i.e., the relative sliding between myofilaments [5]. To understand how the heart beats and how blood vessels regulate their tones, it is important to look into the subcellular structure of these tissues (**Figure 1**).

Heartbeat relies on myofibrils, a fiber bundle structure that abounds in cardiomyocyte (**Figure 1c**). When a number of myofibrils are highly aligned, sarcomere, a repeat unit in the myofibril, can be observed under the microscope. Sarcomere is the basic unit for motion. Two most important proteins in the sarcomere are: myosin and actin. A myosin contains the N-terminal globular head domain, the short neck domain, and the long C-terminal coiled-coil tail domain. The globular head works as a specialized adenosine triphosphatase (ATPase), responsible for adenosine triphosphate (ATP) binding, actin binding, and generating force from ATP hydrolysis. The neck domain transduces force generated by the head. And the fibrous tails are bundled together to form the thick filament (**Figure 1g**). Actin, together with troponin and tropomyosin, forms the thin filament. When the two fibers slide toward each other, the overlapped region increases, which is the mechanism of muscle contraction. Similarly, when the fibers slide away from each other, muscles relax (**Figure 1e**).

2.2 Cross-bridge cycling

The filament sliding depends on cross-bridge cycling [6]. A cross-bridge refers to the two globular heads of myosin, which take turns to bind, pull, and detach from the actin fiber to achieve relative movement between the filaments. An analogy is that a person alternately uses two hands to pull a rope. One alteration of the hand is considered to be one cycle. There are four basic states [7] (sometimes detailed to six states [8]) in cross-bridge cycling. Each state corresponds to one behavior of ATP and one response of myosin. State 1: activation of myosin head, when ATP binds to myosin, it is hydrolyzed to ADP and Pi (inorganic phosphate); myosin becomes the "cocked position." State 2: cross-bridge formation, the activated myosin binds to actin; Pi is released to stabilize the binding. State 3: power stroke, ADP is released; myosin generates force to pull actin filament. State 4: detachment of cross bridge, another ATP binds to myosin; myosin disengages from actin; then State 1 is repeated. The continuous cross-bridge cycling allows myosin to pull actin to its tail side, resulting in

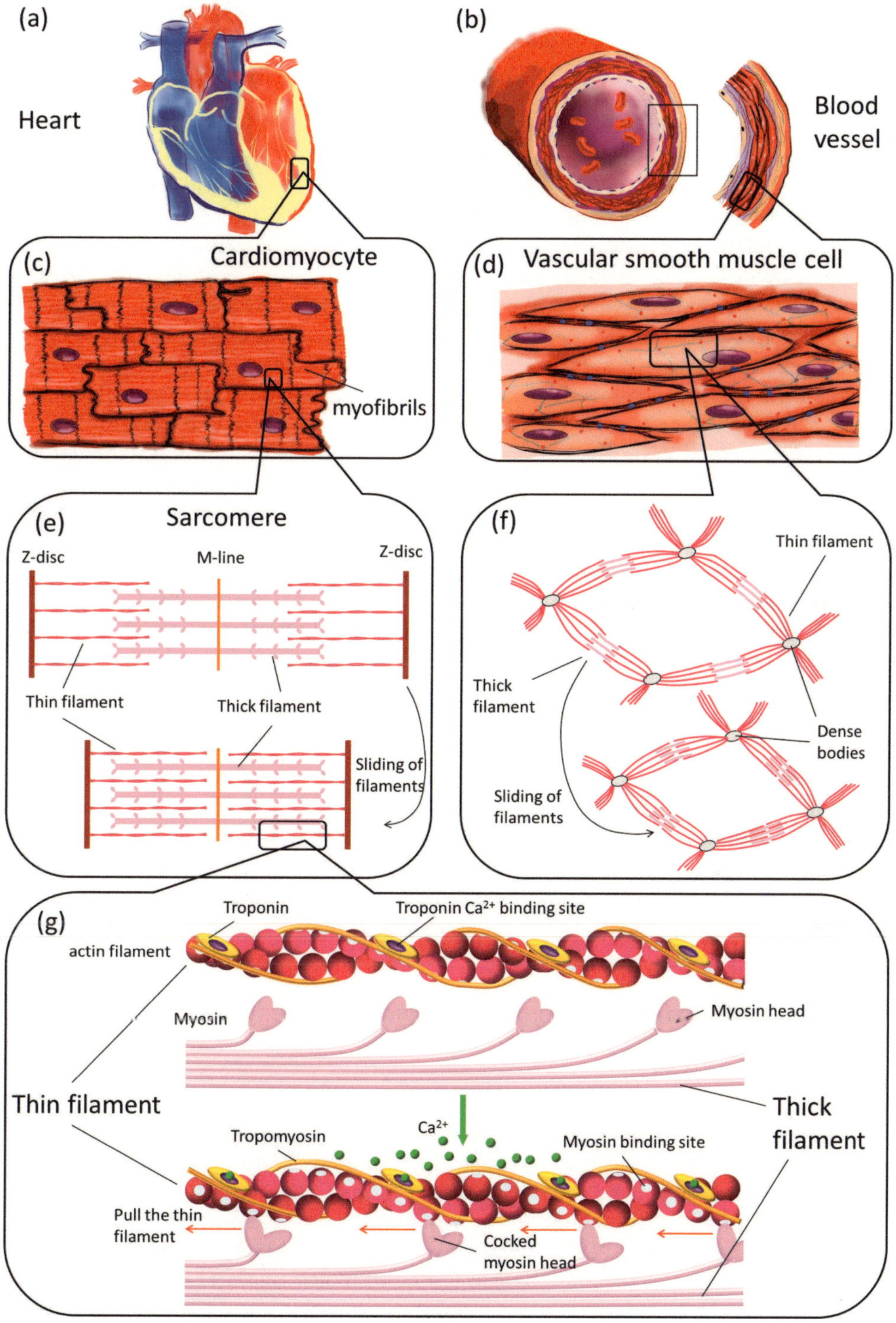

Figure 1.
Muscle contraction illustrated on different structural levels in the cardiovascular system: tissue level—heart (a) and blood vessel (b); cell level—cardiomyocyte (c) and vascular smooth muscle cells (SMCs) (d); subcellular level—filament sliding in cardiomyocyte (e) and SMCs (f); and molecular level—thin and thick filament for cross-bridge cycling (g).

filament sliding and muscle contraction. At the resting state, actin's myosin-binding site is blocked by troponin and tropomyosin [9] (**Figure 1g**). A switch mechanism is needed to expose and mask the myosin-binding site to regulate muscle contraction.

Intracellular Ca^{2+} works as a secondary messenger that quickly bonds to troponin, causing a quick conformational change of troponin and tropomyosin [10]. Thus, the myosin-binding site on actin filaments is exposed, and cross-bridge cycling proceeds. Intracellular Ca^{2+} concentration, or $[Ca^{2+}]_i$, can return to a very low level to cease the contraction and cause relaxation by different mechanisms, such as extruding Ca^{2+} out of cells or storing cytosolic Ca^{2+} into sarcoplasmic reticulum (SR) which functions as the Ca^{2+} reservoir in the cardiomyocyte. Similar mechanisms exist in the vascular tissue. SMC layer lies in between the endothelium layer and adventitia. There is no organized contractile protein fibril or sarcomere structure in SMCs [11]. Instead, the contractile fibrous proteins along with other intermediate filaments form bundles that are immobilized by anchoring proteins onto cell cytoskeletons. These filaments distribute all over the cytoplasm and connect each other through anchoring proteins (dense bodies) to form a three-dimensional network (**Figure 1d** and **f**). Unlike in cardiac muscles, actin filament in smooth muscles is associated with caldesmon, tropomyosin, and calmodulin (CaM) [12]. CaM is an important Ca^{2+} sensing protein, which binds and mediates many enzymes' activities upon Ca^{2+} signaling. Caldesmon binds to actin, which inhibits the activity and motility of actin-myosin ATPase, and this binding is greatly strengthened by tropomyosin [13]. Ca^{2+} binds and activates CaM to uncouple the interaction between caldesmon and actin. Thus, actin's myosin-binding sites are exposed to myosins. Different from skeleton or cardiac tissues, the contraction in smooth muscles also depends on phosphorylation level of myosin light chain, which is adjusted by the enzyme activity of CaM-dependent myosin light chain kinase (CaM-dependent MLCK) and myosin light chain phosphatase (MLCP) [14]. MLCK adds the phosphoryl group to the myosin light chain, while MLCP removes it. Thus, increase of $[Ca^{2+}]_i$ also facilitates muscle contraction through enhancing myosin phosphorylation [15].

Cardiovascular contractility is crucial for blood pressure homeostasis, thermal exchanging, mass transfer, immune responses, and organ functions [16]. Impaired coronary artery contractility (caused by the block of blood vessel or poor dilation) incurs ischemia heart disease [17]. To maintain cardiomyocyte viability and vascular tones, enhancing vasodilation and restricting oxidative stress are critical.

3. NO-related cardiovascular physiology

3.1 Biosynthesis of NO

The biochemical history of NO dates back to 1860s when the therapeutic use of nitroglycerin became a prevalent treatment for angina and hypertension [18]. Consequent studies showed that relaxation of blood vessels depended on a molecule present in an intact endothelium lining, which was named as endothelium-derived relaxing factor (EDRF). In 1977, it was demonstrated that NO was the chemical released from nitroglycerin metabolism [19], and EDRF was later identified to be NO [20]. Since then, NO's cardiovascular effect and medical applications have drawn great research interest.

NO biosynthesis primarily relies on the enzyme nitric oxide synthase (NOS). The reaction uses substrate L-arginine and oxygen to generate L-citrulline and NO in the presence of the cofactors, including Ca^{2+}/CaM, reduced nicotinamide-adenine-dinucleotide phosphate (NADPH), flavin adenine dinucleotide (FAD), flavin mononucleotide (FMN), and tetrahydrobiopterin (BH_4) [21].

NO is an excellent messenger molecule in the human body because it is a highly reactive free radical and a highly diffusible molecule [22]. NO is able to react with various species, such as oxygen, superoxide, lipid oxygen radicals, thiols, and metals [23], which is the biochemical base for NO signaling. The ultimate biological effect of NO depends on its concentration, duration of release, and physiological environment [24]. NO concentration varies from subnanomolar (cell survival signaling) to micromolar (cytotoxic effect and apoptotic signaling) in the body [25]. Its small size and lipophilic characteristic allows it to move rapidly across cell membranes. However, NO's effective distance is limited by its extremely high reactivity with species like oxygen, superoxide, and hemoglobin [23]. NO concentration decreases rapidly around the NO source, which makes NO's effect extremely localized (within hundreds μm).

Three different isoforms of NOS have been identified in the human body. Endothelial NOS (eNOS) is mainly found in vascular endothelial cells (ECs) and is involved in regulating vascular tone and blood clotting. It is also present in cardiomyocytes where it mediates contractility of cardiac muscle [26]. Inducible NOS (iNOS) predominates in immune cells to produce large amounts of NO that aids in the host defense mechanisms. Examples include NO produced by macrophages and neutrophils when induced by lipopolysaccharide, interferon-γ, and tumor necrosis factor alpha [24]. Neuronal NOS (nNOS) is mainly identified in nerve cells where NO assists in nerve transmission. It has also been isolated from cardiomyocytes where NO regulates excitation contraction coupling of cardiac muscles [27]. NO generation by iNOS is mainly regulated at the transcription level, while eNOS and nNOS are constitutive NOS isoforms whose activities are Ca^{2+}- and CaM dependent. Shear forces during blood flow stimulate the opening of Ca^{2+} channels on ECs to increase $[Ca^{2+}]_i$, resulting in eNOS activation [28]. NO biosynthesis can be inhibited by various chemicals that selectively bind to NOS with high affinities, such as N^G-monomethyl L-arginine (L-NMMA) and asymmetric dimethyl L-arginine [9].

S-nitrosothiols and nitrite are alternative endogenous sources of NO. S-nitrosothiols decompose to release NO under physiological pH and are formed through the reaction of NO and thiol [29]. Nitrite can be reduced to NO in the body through pathways involving reducing agents and proteins, such as ascorbic acid, thiols, hemoglobin, and myoglobin [30]. These backup NO generation pathways are emphasized during hypoxia and acidosis when oxygen-dependent NOS-mediated NO production is limited.

3.2 NO-cGMP pathway

Endogenous NO has various physiological effects in the body including inhibiting platelet aggregation, regulating SMC proliferation, modulating immune response, participating in neuron signal transmission, and inducing vasodilation [25]. In the vascular system, NO is primarily generated by eNOS in ECs (**Figure 2**). Shear stresses induced by blood flow and chemical stimuli, known as agonists including acetylcholine, bradykinin, adenosine triphosphate, estrogen, and vascular growth factors, are able to activate eNOS [28, 31]. NO is a highly dynamic molecule that diffuses fast in both aqueous and lipid environment (with diffusion coefficient D = 2.07–3.30 × 10^{-5} cm^2/s in water and polymer matrices [32–34] at physiologically relevant temperatures). Thus, NO readily enters the SMCs and binds to the heme moiety of soluble guanylyl cyclase (sGC) to activate sGC, probably the most important protein target of NO. This binding causes the formation of a nitrosyl-heme complex, heme conformational change, and breaking-apart of sGC His105 from heme iron [35]. Then, an open central core is formed in porphyrin, which can accept guanosine triphosphate (GTP) to generate cyclic guanosine monophosphate (cGMP).

Cyclic-GMP binds and activates cGMP-dependent protein kinase (protein kinase G, or PKG), which can phosphorylate the downstream targets to trigger vasodilation

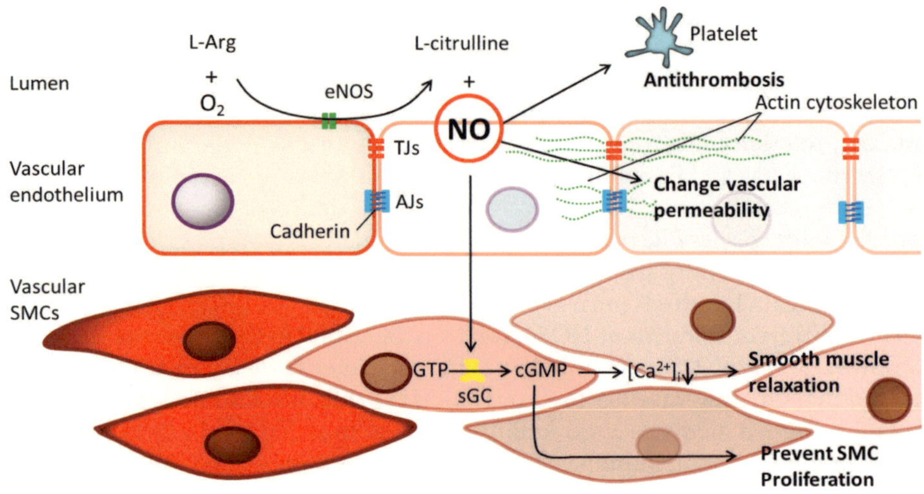

Figure 2.
Biosynthesis of NO by eNOS and the biological effects of NO in the vascular system.

(**Figure 2**). This signaling pathway can be terminated by removing cGMP through converting cGMP to GMP by various phosphodiesterases (PDE) [35]. There are many types of PDE in the human body, and they play critical roles in regulating cardiovascular function, adrenal steroidogenesis, and phototransduction [35].

Interestingly, protein kinase A (PKA, cAMP-dependent protein kinase) and PKG share very similar nucleotide binding domains. Many studies have shown that cGMP can activate PKA downstream pathways and cAMP can also cross-activate PKG [36]. This cross regulation between cGMP and cAMP pathways sometimes complicates NO signaling, which will be shown later.

Due to NO's vasodilation effect, NO releasing drugs such as organic nitrate, and nitro- and nitroso compounds have been used for treating angina pectoris, congestive heart failure damage from ischemia–reperfusion, and pulmonary hypertension [37, 38]. Potent drugs also include chemicals that target members involved in NO-cGMP pathway. For example, Sildenafil, known as Viagra, is a PDE5 inhibitor. It prohibits cGMP from being hydrolyzed by PDE5 and extends the activation time of vasodilation to widen the blood vessel and increases blood flow into the penis to treat erectile dysfunction [39].

3.3 NO-induced muscle relaxation through Ca^{2+} signaling

To fully understand how NO causes vasodilation, it is necessary to perceive the relationship between NO-cGMP pathway and $[Ca^{2+}]_i$. At rest, extracellular Ca^{2+} concentration is high (1–2 mM), while the cytosolic Ca^{2+} is over 1000 times lower (>1 μM) [40, 41]. In the endoplasmic reticulum (ER, or SR in cardiac muscles), Ca^{2+} concentration is also high (about 400 μM) [40]. NO modulates $[Ca^{2+}]_i$ by controlling Ca^{2+} exchange mechanisms on both cell and SR membranes.

3.3.1 Ca^{2+} exchange through plasma membrane

Voltage-gated Ca^{2+} channels regulate $[Ca^{2+}]_i$ through sensing electrical signals to allow Ca^{2+} entering the cell. High voltage-activated L-type channels are broadly found in the cardiovascular system. L-type Ca^{2+} channel inhibitors, such as dihydropyridines, phenylalkylamines, and benzothiazepines, are a major class of drugs for treating CVDs [42]. The opening probability of L-type Ca^{2+} channel can be lowered by PKG indirectly [43]. PKG phosphorylates the K^+ channel and increases its opening probability to hyperpolarize the cell membrane [44, 45]. The hyperpolarized cell membrane can no longer send electrical signals to activate the Ca^{2+} channel, and therefore, Ca^{2+} influx is inhibited. Besides PKG, high NO level also directly activates K^+ channels to achieve aorta relaxation in a cGMP-independent fashion [46].

Ca^{2+} pumping ATPase located on the cell membrane also extrudes Ca^{2+} from the cytosol. It binds Ca^{2+} with a high affinity and forces Ca^{2+} out of the cell even when $[Ca^{2+}]_i$ is very low to maintain the low $[Ca^{2+}]_i$ level at rest [47]. PKG can stimulate the Ca^{2+} pump, initiating the expulsion of cytosolic Ca^{2+} [48].

Unlike Ca^{2+} pump, Na^+/Ca^{2+} exchanger is more effective in quickly removing cytosolic Ca^{2+}, but its Ca^{2+} binding affinity is low [47]. This mechanism is crucial for preventing cells from the cytotoxicity of an acute high Ca^{2+} concentration. The driving force for Na^+/Ca^{2+} exchanger is the stored sodium electrochemical gradient created by Na^+/K^+ channels. PKG can activate the Na^+/K^+ channel to cause more Na^+ accumulated to indirectly facilitate Ca^{2+} removal [49, 50].

3.3.2 Ca^{2+} exchange through ER

Ca^{2+} pump ATPase also resides on the ER responsible for the uptake of cytosolic Ca^{2+} into the ER. NO pathway regulates ER Ca^{2+} pumping through phosphorylation of phospholamban by PKG [51]. Mainly identified in cardiac tissues, phospholamban is an inhibitor of SR Ca^{2+} pump. Phospholamban is normally phosphorylated by PKA, which diminishes its inhibitory effect to Ca^{2+} pump [52]. Interestingly, in neonatal cardiomyocytes and vascular SMCs, NO pathway also demonstrated relaxation effect through differentially phosphorylating phospholamban [53, 54].

Inosital 1,4,5-trisphosphate (IP_3) is a critical messenger molecule that can induce Ca^{2+} release from the ER reservoir. IP_3 receptor resides on the ER and works as a chemical-activated Ca^{2+} channel. NO-cGMP pathway can reduce IP_3 generation [55], and PKG can phosphorylate and inactivate IP_3 receptor in vascular SMCs to inhibit ER Ca^{2+} release [35, 56].

3.3.3 Ca^{2+}-independent muscle relaxation regulated by NO

Independent from NO Ca^{2+} pathway, in SMCs NO also increases MLCP activity and limits MLCK activity, resulting in a dephosphorylation shift of myosin light chain phosphorylation balance [15]. Thus, myosin cross-bridge cycling is inhibited, causing smooth muscle relaxation.

4. Vascular structural integrity mediated by NO

Anatomic alterations in the cardiovascular structure directly deteriorate cardiovascular functions. NO is a multifunctional regulator for homeostasis in the cardiovascular system. An intact endothelial layer is the hub for NO generation. Pathological changes in NO generation can trigger various local flaws that may progress to be systematic cardiovascular issues with time.

4.1 NO-induced alterations in endothelial permeability

Deviant NO level causes change of endothelial permeability, a key characteristic for mass transfer and extravasation. Interestingly, increase, decrease, and no change of vascular permeability due to the presence of NO have been reported. Using high concentration (millimolar level) of exogenous NO donor spermine NONOate decreased endothelial permeability in the *in vitro* human umbilical vein endothelial cell (HUVEC) model [57]. And this effect was amplified by vitamin C, a chemical that increases the apparent half-life of NO. However, in the frog mesenteric capillary model, inhibition of NO synthesis by L-NMMA decreased capillary permeability [58]. Moreover, although NO effectively regulated basal vascular tone in the blood-brain barrier, it demonstrated no effect on its basal permeability [59]. Again, these results demonstrate that NO's biological effect is sensitive to NO concentration, duration, and environment.

Vascular permeability is mainly determined by tightness of cell-cell junctions [60]. Tight junctions (TJs) and adherens junctions (AJs) are the most abundant interendothelial junctions. And both junctions are closely related to actin cytoskeleton dynamics [61] (**Figure 2**). TJs are composed of series of transmembrane proteins that anchor to the actin cytoskeleton to hold cells together. They seal the cells to maintain cell polarity and prevent the molecules from traveling through the space between cells. AJs consist of clusters of transmembrane protein cadherin, which is connected to actin cytoskeleton on its cytoplasmic side and binds strongly with cadherins residing on the neighboring cell membrane. These junctions are important for transmitting mechanical force between cells and reinforcing tissues. Since both junctions directly connect cytoskeletons, the cytoskeleton's behavior will influence cell-cell junctions and thus control vascular permeability. When actin and myosin filaments undergo relative sliding to cause cell contraction, the cytoskeleton-associated membrane proteins will be pulled into the cells, and cell-cell junctions are disrupted. NO mediates cell contraction by adjusting $[Ca^{2+}]_i$. Therefore, deviated NO level may cause the change of cell-cell junctions [60].

Another important downstream molecule of NO is vascular endothelial growth factor (VEGF) which has been extensively studied in cancer research due to its angiogenic effect. VEGF was initially considered as a vascular permeability factor, because it caused the formation of leaky capillaries [62], which is an important characteristic in tumor and retinopathy. Low NO level induces VEGF synthesis under normoxia through the transcription factor hypoxia-inducible factor 1 (HIF-1) [25, 63]. VEGF activates Src kinases, which further phosphorylate cadherin and elicit its internalization [64]. Once cells lose cadherin interactions, gaps between cells form and endothelial permeability is increased.

4.2 Inhibition of SMCs proliferation by NO

One distinctive characteristic of vascular SMCs is the phenotypic plasticity. Two most important phenotypes are contractile and synthetic. Contractile SMCs guarantee the good performance of muscle contraction/relaxation, while synthetic SMCs are highly proliferative and migratory, crucial for vascular remodeling during pregnancy and injury healing. Dysregulation of the phenotype transition causes neointima formation [65]. NO plays important roles in suppressing SMCs' contractile to synthetic transition.

NO donors and 8-Br-cGMP showed similar effect in inhibiting SMC migration and proliferation, indicating NO's inhibitory effect might be through the cGMP-dependent pathway [66]. SMCs overgrow when stimulated by serum and epidermal growth factor (EGF). Many studies were based on these models, though divergent results were reported. EGF induces SMC proliferation through mitogen-activated protein kinases (MAPK) pathway, also called extracellular signal-regulated kinases (ERK) pathway. Ras (a small GTPase) and Raf (kinase of MAPK kinase, or MAPKKK) are the critical upstream protein kinases in this pathway. NO blocks MAPK pathway by prohibiting Raf from being activated by Ras-GTP in rat aortic SMCs. It is believed that PKG phosphorylates Raf, resulting in the conformation change. Thus, Ras-GTP cannot recognize Raf, causing the block of MAPK pathway and the accumulation of Ras-GTP [67]. Meanwhile, elevation of cGMP induced by IL-1β is correlated with the activation of PKA, and it can be prevented by blocking NO and cGMP pathways. Interestingly, this effect is cAMP independent, but PKA inhibitor, not PKG inhibitor, can prevent the inhibition of the proliferation, indicating that cGMP-PKA cross talk plays important roles in suppressing rat aortic SMCs' proliferation [68].

NO-cGMP pathway may inhibit SMC growth by impairing cytoskeleton reorganization. Vasodilator-stimulated phosphoprotein (VASP) is characterized as a substrate of both PKG and PKA [69]. It targets focal adhesions and is involved in actin filament formation. Cell morphology change during proliferation relies on VASP, and its activation relies on the phosphorylation of Ser157 primarily mediated by PKA [70]. However, PKG can phosphorylate Ser239 and Thr278 to impair VASP's activity and inhibit actin cytoskeleton reorganization [70, 71].

NO also directly mediates proteins associated with cell cycle and cell metabolism by cGMP-independent mechanisms. Cyclin A and cyclin-dependent kinase 2 expression levels can be blunted by exogenous NO donor DETA NONOate in an *in vitro* vascular SMC model [72]. Ornithine decarboxylase (ODC) catalyzes the ornithine decarboxylation to form polyamines, which are necessary for cell growth and proliferation. ODC's active center can be masked by nitrosylation. And NO biosynthesis's intermediate product N(omega)-hydroxyarginine can inhibit ODC enzyme activity [73] to disrupt cell proliferation.

4.3 Prevention of thrombogenesis by NO

Thrombus formation is critical for hemostasis during injury. However, thrombus in blood vessels can cause stroke and heart attack. Stable thrombus reduces lumen size and stiffens blood vessels. Unstable thrombus may rupture with blood flow and block the vessel. Activation of platelet is a critical step for thrombus formation, which involves exocytosis processes to expose P-selectin on the platelet surface and activate glycoprotein IIb/IIIa. Both processes depend on the elevation of $[Ca^{2+}]_i$ controlled by IP_3 pathway. NO suppresses platelet activation through NO-cGMP pathway [74]. Although, the inhibition pathway has not been fully characterized, evidences have shown that cGMP-PKG blocks agonist-induced IP_3 formation in platelet [75], and PKG can phosphorylate IP_3 receptor to inhibit Ca^{2+} release from the ER [35].

When the endothelium loses its integrity, there will be a local shortage of thromboregulators such as NO, prostacyclin, and ectonucleotidase CD 39, resulting in thrombogenesis [76]. Collagen and tissue factors also trigger the coagulation reactions [76]. The use of blood contact implant is another common source of thrombus. Note that, all materials are thrombogenic to some degrees. To enhance implant biocompatibility, an efficient method is to use NO releasing polymers to fabricate or surface coat the blood contacting devices (such as vascular graft/stent, intravascular catheter, and sensor implants). Common strategies include: physically incorporating NO releasing chemicals into polymer matrices, chemically linking NO releasing agent to polymer backbones, and developing materials that can trigger NO generation using endogenous NO donors circulating in the blood. By using the first two strategies, successful trials have been reported to achieve long-term (over few weeks to months) NO releasing and antithrombotic applications [77–79]. Good NO donors include N-diazeniumdiolate and S-nitrosothiols. Both hydrophilic and hydrophobic polymers that are commonly used in medical device fabrication have been successfully modified for NO release including poly(vinyl chloride), polymethacrylates, various hydrogels, polyethylene terephthalate, polyurethane, and silicone rubbers [77]. The third strategy directly uses endogenous NO donors as the NO reservoir to catalyze NO generation from S-nitrosoglutathione or nitrite in the body. Currently, its main challenge is to adjust the NO releasing rate to be more biologically relevant.

5. Nitroso-redox balance in the cardiovascular system

Oxidative stress is always associated with ischemia reperfusion injury, dilated cardiomyopathy, and heart failure [80]. It is crucial to restore redox balance in the cardiovascular system when treating these diseases. Redox balance is governed by changes in the oxidative state in tissues, where addition and loss of electrons result in reduction and oxidation of molecules, respectively [80]. Oxygen can accept an electron to become reactive oxygen species (ROS). ROS are highly reactive chemical species that contain oxygen atoms, mostly free radicals with one or more unpaired electrons [81]. NO is a free radical signaling molecule. Under pathological conditions, it reacts with superoxide to generate reactive nitrogen species (RNS) that have detrimental consequences to cells. Herein, we highlight the causes of redox imbalance, their functions in the cardiovascular system, and the roles they play in the progression of CVDs.

5.1 Biochemistry and physiology of ROS and RNS

5.1.1 ROS and oxidative stress

The electron transport chain (ETC) located in the inner membrane of mitochondria is crucial for energy and ROS generation (**Figure 3**). Normally, the final electron acceptor oxygen is reduced to water. However, in pathological conditions, electrons uncouple from the chain and react with oxygen without passing *cytochrome c* oxidase to form superoxide. Other ROS sources include NADPH oxidase, xanthine oxidase (XO), eNOS, and cytochrome P450s (CYP). NADPH oxidases belong to NOX family proteins, which transfer electrons across intracellular membranes. NADPH oxidases transfer electron from NADPH to oxygen to form superoxide for immune responses [82, 83]. Three NOX enzymes have been found in the vascular wall, NOX 1, 2, and 4. NOX 1 and 2 result in the formation of superoxide and NOX 4 produces hydrogen peroxide (H_2O_2) [83]. NOX 4 is also present in the mitochondria and SR in cardiomyocytes [80]. XO is found in the heart and ECs. It catalyzes purine metabolism with superoxide and H_2O_2 produced. The activity of XO is enhanced under ischemic reperfusion injury and oscillatory shear stress [83, 84]. Another important source of ROS is the uncoupling of eNOS, which causes eNOS to produce superoxide, instead of NO. One of the reasons for eNOS uncoupling is the deficiency in the substrates (L-arginine and oxygen) and co-factor BH_4 [80, 83]. Uncoupling of eNOS may explain why diabetic patients are susceptible to CVDs. High glucose increases arginase levels in ECs, which competes with eNOS for the substrate L-arginine [85]. In addition, overload of ROS in diabetes mellitus limits BH_4 biosynthesis to further facilitate eNOS uncoupling [86]. Cytochrome p450s (CYP) are a group of hemoproteins similar in structure and function to eNOS under oxidative stresses. The catalytic activity of CYP requires oxygen and two electrons to form a ferrous-dioxy complex [87]. CYP is involved in the metabolism of cholesterol, vitamins, and arachidonic acid [88]. When there is excessive oxygen consumed, the enzyme uncouples, and the ferrous-dioxy complex diverts back to the ferric state to produce superoxide [89].

Superoxide is an anion-free radical that can produce other ROS including H_2O_2, hydroxyl radicals ($^{\cdot}OH$), and hypochlorous acid (HClO) [80, 82, 90, 91].

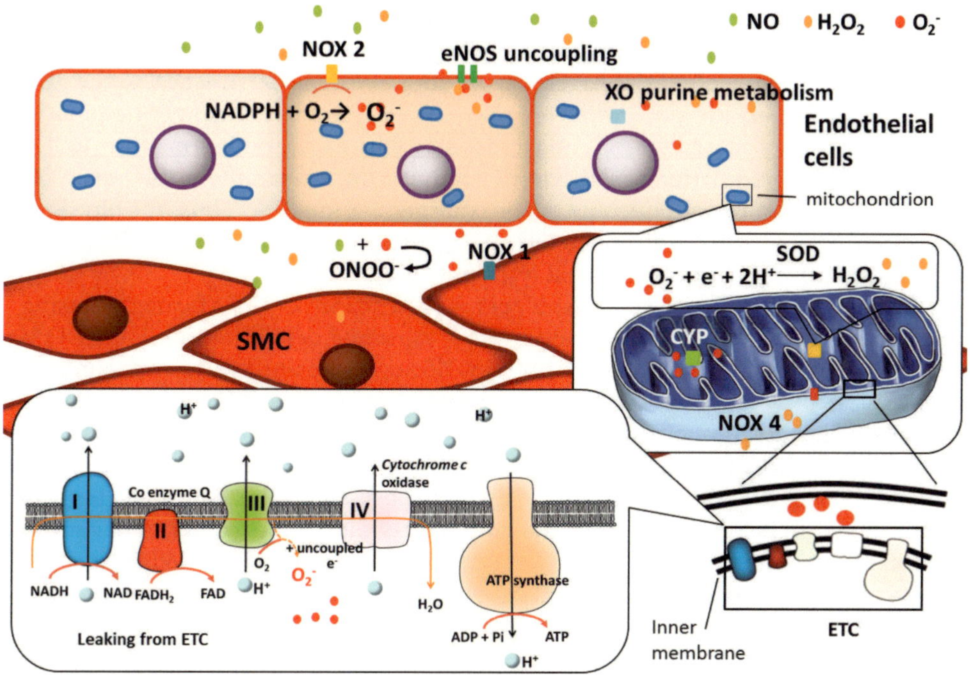

Figure 3.
The major sources of ROS and RNS in the cardiovascular system.

The spontaneous transfer of an electron to superoxide at low pH or by an enzyme reaction (superoxide dismutase, SOD) produces H_2O_2 [82, 91]. Low levels of H_2O_2 (1–10 nM) induce more antioxidant molecules that protect the cells, and high levels (>100 nM) are likely to generate more prooxidants [91]. For example, high level of H_2O_2 was generated in neutrophils for antimicrobial effects [92, 93]. Hydroxyl radical can be formed from the reaction between H_2O_2 and superoxide (Haber Weiss reaction) or the breakdown of H_2O_2 by metal ions, Fe^{2+} or Cu^{2+} (Fenton reaction) [94]. Hydroxyl radical is highly reactive. It alters DNA structure by attacking purine and pyrimidine bases, leading to mutations and cell damages [95]. In the pathological myocardial tissue, it is associated with decreased contractile function, increased membrane phospholipid peroxidation, and heart failure [96, 97]. HClO is mainly produced by leukocytes when H_2O_2 reacts with chloride anions. It facilitates the removal of foreign particles and is also implicated in the progression of atherosclerosis and ischemic reperfusion injury [81].

5.1.2 NO and nitrosative stress

NO acts in a diffusion- and concentration-dependent manner. Low concentrations of NO (nanomolar range) have a protective role, while high NO levels (micromolar range) can be detrimental [98]. The majority of NO's biological effect is attributed to sGC/cGMP pathway [21]. Additionally, NO acts as a signaling mediator through S-nitrosylation. NO can inhibit cardiac hypertrophy through nitrosylation of histone deacetylase 2 (HDAC2) released from chromatin [99]. HDAC2 regulates anti-hypertrophic genes. In ischemic preconditioning (the body's defense mechanism against myocardial necrosis), the S-nitrosylation of mitochondrial proteins protects the mitochondria from oxidative stress [100]. S-nitrosylation initiates excitation contraction coupling by increasing Ca^{2+} uptake, and the contraction may be sustained through releasing of Ca^{2+} via SR membrane ryanodine receptors

(RyRs) [26]. Quantitatively, when three thiols per subunit of RyR channels are nitrosylated, the process is reversible. However, if six or more thiols per subunit are nitrosylated, irreversible Ca^{2+} ion release occurs and can be detrimental to the cardiac muscle [26]. In addition, when too much NO is generated during inflammation or sepsis, NO may cause hypovolemia due to its excessive vasodilation effect [83]. Furthermore, upregulation of iNOS in ECs reduces the availability of BH_4 to eNOS, intensifying eNOS uncoupling and superoxide generation [83]. Thus, the physiological role of NO can be attenuated by ROS, because NO is quickly consumed by superoxide before it initiates any cell response [101].

When NO collides with superoxide, peroxynitrite ($ONOO^-$) is formed, causing nitrosative stress. The chemical reaction is very fast and deleterious [98]. $ONOO^-$ is a very strong oxidant. It reacts with proteins through tyrosine and tryptophan residues to form nitrotyrosine and nitrotryptophan, respectively [80, 98]. In diabetic mice, tyrosine nitration of the voltage-gated K^+ channels in the vascular SMCs altered its dilation function, a possible mechanism of the progression of coronary artery disease [102]. Tyrosine nitration was also observed in cardiac myocytes desmin, myosin heavy chain, α-actin, and microtubules. These proteins play pivotal roles in maintaining cell morphology and cardiac contractility [98]. When free nitrotyrosine was incorporated into the carboxyl terminus of α-tubulin in microtubules, altered microtubule organization and redistribution of the motor cytoplasmic protein dynein were observed [103]. Protein activity can also be impaired by oxidation of thiols to disulfide bond by $ONOO^-$ [98]. In addition, $ONOO^-$ also reacts with lipids to yield nitrated lipids to promote atherosclerosis, and with nucleic acids via guanine and the sugar phosphate backbone to damage DNA [98].

On the other hand, low concentrations of $ONOO^-$ (10–200 uM) is associated with tyrosine kinase-dependent signaling. $ONOO^-$ has been shown to activate tyrosine phosphorylation and trigger glycolysis [98]. Another example involves MAPK pathway, where $ONOO^-$ activates Raf-1 kinase. The MAPK pathway is closely associated with anti-apoptosis and cardiac hypertrophy in the cultured cardiomyocyte model [104].

5.2 Atherosclerosis

5.2.1 Inflammatory mechanism of atherosclerosis

Atherosclerosis is characterized by the formation of plaques that reduce the lumen of arteries and consequently interfere with blood flow and tissue perfusion. The plaque consists of the lipid core and fibrous cap. In patients with hypercholesterolemia, ROS and RNS oxidize low-density lipoprotein (LDL) [105]. Oxidized LDL (Ox LDL) initiates a cascade of events that alters the endothelial permeability and leads to insudation of the lipoprotein in the arterial wall. Stimulated by atheroprone signals, ECs express selectins and vascular cell adhesion molecule (VCAM-1) to attract circulating blood monocytes. Monocytes penetrate the endothelial layer; i.e., diapedesis occurs, and become macrophages [106]. Macrophages target Ox LDL for phagocytosis and become foam cells, the accumulation of which causes the formation of fatty streaks. The foam cells initiate the production of transforming growth factor beta (TGF-β), platelet-derived growth factor (PDGF), and fibroblast growth factor (FGF) in the vascular system [107, 108]. These growth factors promote the change of vascular SMCs from a contractile to a synthetic phenotype. SMCs migrate from the media layer to the intima, where they secrete a complex extracellular matrix to form a fibrous cap around the lipid core to stabilize the plaque [109]. The proliferation of SMCs leads to neointima hyperplasia. Thus, the vessel becomes narrowed and the blood flow profile alters, further aggravating endothelial dysfunction (**Figure 4**).

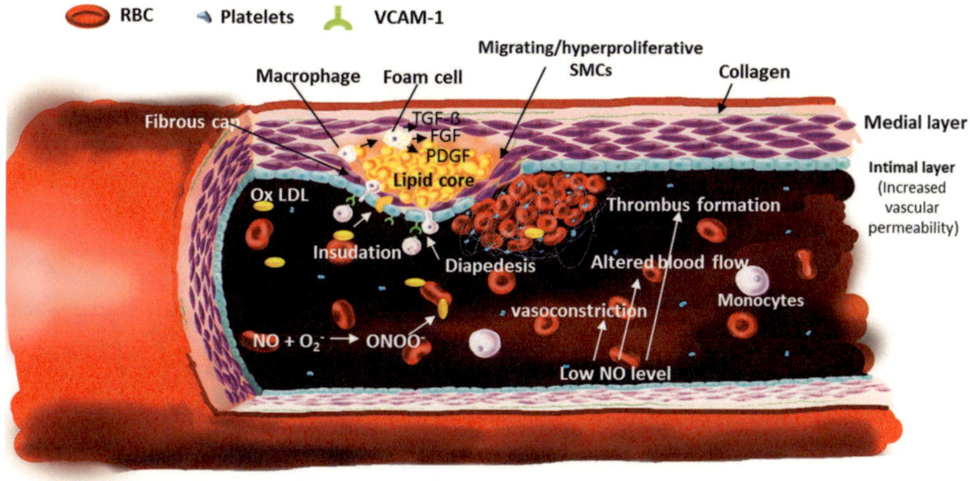

Figure 4.
NO's role in the initiation and progression of atherosclerosis.

5.2.2 Hemodynamics and atherosclerosis

In fact, disturbed blood flow at arches, branches, or bifurcations is always associated with the early appearance and fast development of atherosclerotic lesions. Blood flow influences ECs' gene expression through "shear-stress response elements" in the promoters of atherosclerosis relevant genes and "mechanotransducers" that can sense the force and transduce mechanical signal into biochemical events within the cell. Overall, in steady laminar flow, ECs express more antithrombotic, anti-inflammatory, and antioxidant proteins, such as eNOS, cyclooxygenase-2 (COX-2), and manganese-dependent superoxide dismutase (SOD) [110], while in turbulent flow, ECs show atheroprone phenotypes, which activate NF-κB pathways to promote the expression of cytokines and cell adhesion molecules [107].

Two highly differentially expressed transcription factors, zinc finger transcription factor Kruppel-like factor 2 (KLF2) and nuclear factor erythroid-2-related factor-2 (Nrf2), were identified by comparing endothelial gene expressions under different hemodynamic patterns [111]. KLF-2 maintains endothelial homeostasis at least in part by inhibiting local inflammation and restoring NO levels. Overexpression of KLF-2 blocks IL-1β-induced inflammation through inhibiting VCAM-1 and E-selectin expression to disturb the adhesion of immune cells [112]. In addition, it upregulates eNOS expression to improve vascular tones. Nrf2 is responsible for regulating redox-related genes (heme oxygenase 1, ferritin heavy chain, NADPH dehydrogenase quinone 1, and thioredoxin reductase) to maintain vascular redox balance in laminar flow [111]. Remarkably, it has been shown that KLF2 and Nrf2 work synergistically to integrate atheroprotective signals and active antioxidant responses, which may be a promising therapeutic strategy for CVDs.

5.3 Antioxidant mechanisms in nitroso-redox balance

To counteract the effect of excessive ROS and control CVD symptoms, introducing antioxidative mechanisms is an effective method (**Figure 5**). Increasing enzymes that can eliminate ROS is a commonly used strategy. For example, superoxide can be eliminated by dismutation of two superoxide molecules by SOD to O_2 and H_2O_2 [113]. H_2O_2 can undergo decomposition under the regulation of catalase and peroxiredoxin to oxygen and water [80, 114]. The thiol group in peroxiredoxins

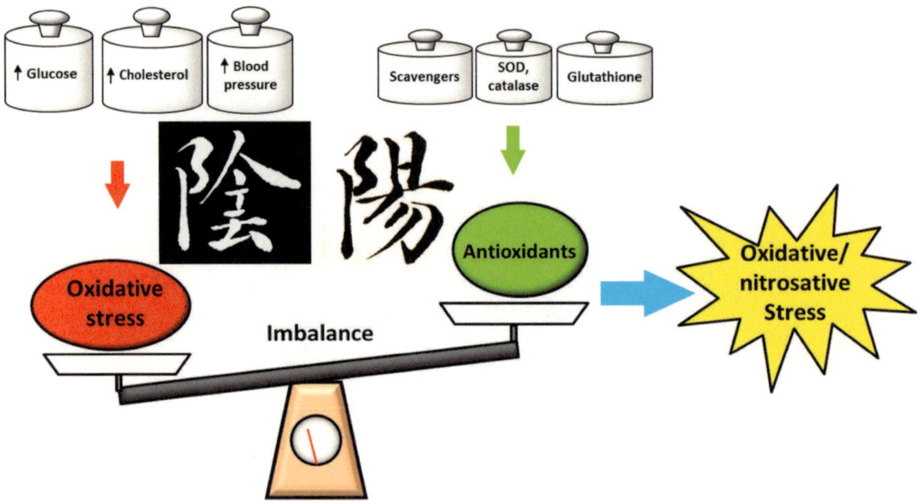

Figure 5.
Maintenance of redox balance in the cardiovascular system.

consumes H_2O_2 to form sulfenic acid, then subsequently disulfide bond [115]. Glutathione (GSH) peroxidase 1 uses the similar mechanism to inactivate H_2O_2, superoxide, and $ONOO^-$ in the presence of the tripeptide compound GSH. A prospective cohort study showed that reduced levels of GSH peroxidase 1 were associated with increased mortality in coronary disease patients [116].

Another effective antioxidative method is to protect redox-sensitive molecules from being oxidized. In the body, the thiol group on GSH can form reversible mixed disulfide bonds with cellular proteins under oxidative stress conditions. These disulfide bonds can be broken by the enzyme glutaredoxin when the surrounding cell environment reverts back to its normal state [80, 117]. The addition of scavengers to directly remove ROS/RNS can also restore the nitroso-redox balance. An example is the elimination of superoxide by ascorbic acid (vitamin C) [113]. By limiting superoxide, other reactive species can also be repressed, such as $^{\bullet}OH$ and $ONOO^-$. This may explain the success of the clinical trial of combining nitrate drug isosorbide dinitrate with hydralazine, a NADPH oxidase inhibitor, where heart failure was reduced by 45% [118]. By inhibiting superoxide generation from NADPH oxidase, $ONOO^-$ level may be reduced and NO function preserved.

The high concentrations of NO can be controlled through scavenging NO via oxyhemoglobin in red blood cells and myoglobin in the skeletal and heart muscle. These two proteins react with NO to form nitrate, which is considered as the primary method for inactivating NO in the cardiovascular system [119]. Hemoglobin and myoglobin can also scavenge $ONOO^-$ by their metal centers, generating nitrate from the reactions [120].

6. Conclusions and future outlooks

We briefly reviewed the molecular mechanisms of muscle contraction and relaxation in the cardiovascular system and highlighted the importance of physiological and pathological effects of NO and oxidative stress. NO and ROS both determine the structural integrity and functionality of the cardiovascular system. The cardiovascular system not only nourishes cells, but also provides paths for immune response and systematic signaling. Drugs are transported by this system to the correct site for metabolic reactions. Tissue regeneration also relies on a healthy

DOI: http://dx.doi.org/10.5772/intechopen.82556

cardiovascular system. Therefore, to maintain, the homeostasis of the cardiovascular system is essential for overall health. Unfortunately, with aging, both cardiac function and cardiomyocyte number decline [121], and blood vessels undergo structural alterations [122]. Moreover, CVDs are also associated with other serious complications, such as diabetes, cancer, kidney failure, and inflammatory processes. Thus, multiple therapeutic strategies are needed to treat CVDs. According to 2011's American Heart Association's guidelines for preventing CVDs, therapeutic strategies include smoking cessation, blood pressure control, lipid management, physical activity programs, diabetes management, anticoagulation, dilation management, and depression prevention [3]. Besides traditional pharmaceutical management and surgeries, new perspectives to study, diagnose, and treat CVDs have also shown promising results, including development of biocompatible stents [123], stem cells therapies [124, 125], novel devices for mechanical thrombectomy [126], and inflammation management [127]. Although challenges still exist, the implementations of research findings from different disciplines in clinical trials will allow us to better understand and control CVDs in the future.

Conflict of interest

The authors have declared that no conflict of interest exists.

Author details

Weilue He[1]*, Maria Paula Kwesiga[1], Eyerusalem Gebreyesus[1] and Sijia Liu[2]

1 Department of Biomedical Engineering, Michigan Technological University, Houghton, MI, USA

2 Rehabilitation Medicine Center, West China Hospital, Sichuan University, Chengdu, China

*Address all correspondence to: weilueh@mtu.edu

References

[1] The Top 10 Causes of Death [World Health Organization]. 2016. Available from: http://www.who.int/news-room/fact-sheets/detail/the-top-10-causes-of-death [Accessed: 11 November 2018]

[2] Dahlöf B. Cardiovascular disease risk factors: Epidemiology and risk assessment. The American journal of cardiology. 2010;**105**:3A-9A. DOI: 10.1016/j.amjcard.2009.10.007

[3] Smith SC et al. Aha/Accf secondary prevention and risk reduction therapy for patients with coronary and other atherosclerotic vascular disease: 2011 update: A Guideline from the American Heart Association and American College of Cardiology Foundation Endorsed by the World Heart Federation and the Preventive Cardiovascular Nurses Association. Journal of the American College of Cardiology. 2011;**58**:2432-2446. DOI: 10.1016/j.jacc.2011.10.824

[4] Taylor AL et al. Combination of isosorbide dinitrate and hydralazine in blacks with heart failure. New England Journal of Medicine. 2004;**351**: 2049-2057. DOI: 10.1056/NEJMoa042934

[5] Huxley HE. The mechanism of muscular contraction. Science. 1969;**164**:1356-1365. DOI: 10.1126/science.164.3886.1356

[6] Eisenberg E, Hill TL, Chen Y-D. Cross-bridge model of muscle contraction. quantitative analysis. Biophysical Journal. 1980;**29**:195-227. DOI: 10.1016/S0006-3495(80)85126-5

[7] Marieb EN, editor. Essentials of Human Anatomy & Physiology. 11th ed. London, England: Pearson Education; 2007

[8] Geeves MA, Fedorov R, Manstein DJ. Molecular mechanism of actomyosin-based motility. Cellular and Molecular Life Sciences. 2005;**62**:1462-1477. DOI: 10.1007/s00018-005-5015-5

[9] Zot AS, Potter JD. Structural aspects of troponin-tropomyosin regulation of skeletal muscle contraction. Annual Review of Biophysics and Biophysical Chemistry. 1987;**16**:535-559. DOI: 10.1146/annurev.bb.16.060187.002535

[10] Lehman W, Craig R, Vibert P. Ca^{2+}-induced tropomyosin movement in limulus thin filaments revealed by three-dimensional reconstruction. Nature. 1994;**368**:65. DOI: 10.1038/368065a0

[11] Arner A, Pfitzer G. Regulation of cross-bridge cycling by Ca^{2+} in smooth muscle. In: Reviews of Physiology Biochemistry and Pharmacology. Berlin Heidelberg: Springer-Verlag; 1999. pp. 63-146. DOI: 10.1007/3-540-64753-8

[12] Takahashi K, Hiwada K, Kokubu T. Isolation and characterization of a 34,000-dalton calmodulin-and F-actin-binding protein from chicken gizzard smooth muscle. Biochemical and Biophysical Research Communications. 1986;**141**:20-26. DOI: 10.1016/S0006-291X(86)80328-X

[13] Sobue K, Sellers JR. Caldesmon, a novel regulatory protein in smooth muscle and nonmuscle actomyosin systems. Journal of Biological Chemistry. 1991;**266**:12115-12118

[14] Horowitz A et al. Mechanisms of smooth muscle contraction. Physiological Reviews. 1996;**76**:967-1003. DOI: 10.1152/physrev.1996.76.4.967

[15] Etter EF et al. Activation of myosin light chain phosphatase in intact arterial smooth muscle during nitric oxide-induced relaxation. Journal of Biological Chemistry. 2001. DOI: 10.1074/jbc.M104737200

[16] Towbin JA, Bowles KR, Bowles NE. Etiologies of cardiomyopathy and heart failure. Nature Medicine. 1999;**5**:266. DOI: 10.1038/6474

[17] Healy B. Endothelial cell dysfunction: An emerging endocrinopathy linked to coronary disease. Journal of the American College of Cardiology. 1990;**16**:357-358. DOI: 10.1016/0735-1097(90)90585-D

[18] Ignarro LJ. Nitric oxide: A unique endogenous signaling molecule in vascular biology. Bioscience Reports. 1999;**19**:51-71. DOI: 10.1023/A:1020150124721

[19] Katsuki S et al. Stimulation of guanylate cyclase by sodium nitroprusside, nitroglycerin and nitric oxide in various tissue preparations and comparison to the effects of sodium azide and hydroxylamine. Journal of Cyclic Nucleotide Research. 1977;**3**:23-35

[20] Ignarro LJ et al. Endothelium-derived relaxing factor produced and released from artery and vein is nitric oxide. Proceedings of the National Academy of Sciences. 1987;**84**: 9265-9269. DOI: 10.1073/pnas.84.24.9265

[21] Murad F. Nitric oxide and cyclic gmp in cell signaling and drug development. New England Journal of Medicine. 2006;**355**:2003-2011. DOI: 10.1056/NEJMsa063904

[22] Loscalzo J, Welch G. Nitric oxide and its role in the cardiovascular system. Progress in Cardiovascular Diseases. 1995;**38**:87-104. DOI: 10.1016/S0033-0620(05)80001-5

[23] Lancaster J Jr. A tutorial on the diffusibility and reactivity of free nitric oxide. Nitric Oxide. 1997;**1**:18-30. DOI: 10.1006/niox.1996.0112

[24] Moncada S, Higgs A. The L-arginine-nitric oxide pathway. New England Journal of Medicine. 1993;**329**:2002-2012. DOI: 10.1056/NEJM199312303292706

[25] Thomas DD et al. The chemical biology of nitric oxide: Implications in cellular signaling. Free Radical Biology and Medicine. 2008;**45**:18-31. DOI: 10.1016/j.freeradbiomed.2008.03.020

[26] Saraiva RM, Hare JM. Nitric oxide signaling in the cardiovascular system: Implications for heart failure. Current Opinion in Cardiology. 2006;**21**:221-228. DOI: 10.1097/01.hco.0000221584.56372.dc

[27] Khan SA et al. Neuronal nitric oxide synthase negatively regulates xanthine oxidoreductase inhibition of cardiac excitation-contraction coupling. Proceedings of the National Academy of Sciences. 2004;**101**:15944-15948. DOI: 10.1073/pnas.0404136101

[28] Kuchan M, Frangos J. Role of calcium and calmodulin in flow-induced nitric oxide production in endothelial cells. American Journal of Physiology-Cell Physiology. 1994;**266**:C628-C636. DOI: 10.1152/ajpcell.1994.266.3.C628

[29] Singh RJ et al. Mechanism of nitric oxide release from s-nitrosothiols. Journal of Biological Chemistry. 1996;**271**:18596-18603. DOI: 10.1074/jbc.271.31.18596

[30] Lundberg JO, Weitzberg E, Gladwin MT. The nitrate–nitrite–nitric oxide pathway in physiology and therapeutics. Nature Reviews Drug Discovery. 2008;**7**:156. DOI: 10.1038/nrd2466

[31] Fulton D et al. Agonist-stimulated endothelial nitric oxide synthase activation and vascular relaxation: Role of enos phosphorylation at Tyr83. Circulation Research. 2008;**102**:497-504. DOI: 10.1161/CIRCRESAHA.107.162933

[32] Wise D, Houghton G. Diffusion coefficients of neon, krypton, xenon, carbon monoxide and nitric oxide in water at 10-60°C. Chemical Engineering Science. 1968;**23**:1211-1216. DOI: 10.1016/0009-2509(68)89029-3

[33] Malinski T et al. Diffusion of nitric oxide in the aorta wall monitored in

situ by porphyrinic microsensors. Biochemical and Biophysical Research Communications. 1993;**193**:1076-1082. DOI: 10.1006/bbrc.1993.1735

[34] Mowery K, Meyerhoff M. The transport of nitric oxide through various polymeric matrices. Polymer. 1999;**40**:6203-6207. DOI: 10.1016/S0032-3861(99)00187-1

[35] Lucas KA et al. Guanylyl cyclases and signaling by cyclic Gmp. Pharmacological Reviews. 2000;**52**:375-414

[36] Jiang H et al. Direct evidence for cross-activation of cgmp-dependent protein kinase by camp in pig coronary arteries. Journal of Biological Chemistry. 1992;**267**:1015-1019

[37] Nossaman VE, Nossaman BD, Kadowitz PJ. Nitrates and nitrites in the treatment of ischemic cardiac disease. Cardiology in Review. 2010;**18**:190. DOI: 10.1097/CRD.0b013e3181c8e14a

[38] Miller M, Megson I. Recent developments in nitric oxide donor drugs. British Journal of Pharmacology. 2007;**151**:305-321. DOI: 10.1038/sj.bjp.0707224

[39] Corbin JD, Francis SH. Cyclic Gmp phosphodiesterase-5: Target of sildenafil. Journal of Biological Chemistry. 1999;**274**:13729-13732. DOI: https://doi.org/10.1074/jbc.274.20.13729

[40] Gogia S et al. Role of calcium in regulating the intra-and extracellular cleavage of Von Willebrand Factor by the protease Adamts13. Blood Advances. 2017;**1**:2063-2074. DOI: 10.1182/bloodadvances.2017009027

[41] Brunet J-P et al. Rotavirus infection induces an increase in intracellular calcium concentration in human intestinal epithelial cells: Role in microvillar actin alteration. Journal of Virology. 2000;**74**:2323-2332. DOI: 10.1128/JVI.74.5.2323-2332.2000

[42] Diochot S et al. Dihydropyridines, phenylalkylamines and benzothiazepines block N-, P/Q-and R-type calcium currents. Pflügers Archiv. 1995;**431**:10-19. DOI: 10.1007/BF00374372

[43] Hofmann F, Ammendola A, Schlossmann J. Rising behind no: Cgmp-dependent protein kinases. Journal of Cell Science. 2000;**113**:1671-1676

[44] Furukawa K et al. Activation of K^+ channels and suppression of neuronal activity by secreted β-amyloid-precursor protein. Nature. 1996;**379**:74. DOI: 10.1038/379074a0

[45] Robertson BE et al. Cgmp-dependent protein kinase activates Ca-activated K channels in cerebral artery smooth muscle cells. American Journal of Physiology-Cell Physiology. 1993;**265**:C299-C303. DOI: 10.1152/ajpcell.1993.265.1.C299

[46] Bolotina VM et al. Nitric oxide directly activates calcium-dependent potassium channels in vascular smooth muscle. Nature. 1994;**368**:850. DOI: 10.1038/368850a0

[47] Brini M, Carafoli E. The plasma membrane Ca^{2+} atpase and the plasma membrane sodium calcium exchanger cooperate in the regulation of cell calcium. Cold Spring Harbor Perspectives in Biology. 2011;**3**:a004168. DOI: 10.1101/cshperspect.a004168

[48] Yoshida Y et al. Cyclic Gmp-dependent protein kinase stimulates the plasma membrane Ca^{2+} pump atpase of vascular smooth muscle via phosphorylation of a 240-Kda protein. Journal of Biological Chemistry. 1991;**266**:19819-19825

[49] Nishimura J. Topics on the Na^+/Ca^{2+} exchanger: Involvement of Na^+/Ca^{2+} exchanger in the vasodilator-induced vasorelaxation. Journal of Pharmacological Sciences.

2006;**102**:27-31. DOI: 10.1254/jphs.FMJ06002X5

[50] Tamaoki J et al. Role of Na^+-K^+ atpase in cyclic Gmp-mediated relaxation of canine pulmonary artery smooth muscle cells. British Journal of Pharmacology. 1997;**122**:112-116. DOI: 10.1038/sj.bjp.0701351

[51] Cornwell TL et al. Regulation of sarcoplasmic reticulum protein phosphorylation by localized cyclic Gmp-dependent protein kinase in vascular smooth muscle cells. Molecular Pharmacology. 1991;**40**:923-931

[52] Masterson LR et al. Camp-dependent protein kinase a selects the excited state of the membrane substrate phospholamban. Journal of Molecular Biology. 2011;**412**:155-164. DOI: 10.1016/j.jmb.2011.06.041

[53] Karczewski P et al. Role of phospholamban in No/Edrf-induced relaxation in rat Aorta. Life Sciences. 1992;**51**:1205-1210. DOI: 10.1016/0024-3205(92)90357-U

[54] Gorbe A et al. Role of Cgmp-Pkg signaling in the protection of neonatal rat cardiac myocytes subjected to simulated ischemia/reoxygenation. Basic Research in Cardiology. 2010;**105**:643-650. DOI: 10.1007/s00395-010-0097-0

[55] Ruth P et al. Transfected Cgmp-dependent protein kinase suppresses calcium transients by inhibition of Inositol 1, 4, 5-trisphosphate production. Proceedings of the National Academy of Sciences. 1993;**90**:2623-2627. DOI: 10.1073/pnas.90.7.2623

[56] Komalavilas P, Lincoln TM. Phosphorylation of the inositol 1, 4, 5-trisphosphate receptor by cyclic Gmp-dependent protein kinase. Journal of Biological Chemistry. 1994;**269**:8701-8707. DOI: 10.1074/jbc.271.36.21933

[57] May JM, Qu Z-C. Nitric oxide mediates tightening of the endothelial barrier by ascorbic acid. Biochemical and Biophysical Research Communications. 2011;**404**:701-705. DOI: 10.1016/j.bbrc.2010.12.046

[58] Rumbaut RE, McKay MK, Huxley VH. Capillary hydraulic conductivity is decreased by nitric oxide synthase inhibition. American Journal of Physiology-Heart and Circulatory Physiology. 1995;**268**:H1856-H1861. DOI: 10.1152/ajpheart.1995.268.5.H1856

[59] Mayhan WG. Inhibition of nitric oxide synthase does not alter basal permeability of the blood–brain barrier. Brain Research. 2000;**855**:143-149. DOI: 10.1016/S0006-8993(99)02389-6

[60] Predescu D et al. Constitutive enos-derived nitric oxide is a determinant of endothelial junctional integrity. American Journal of Physiology-Lung Cellular and Molecular Physiology. 2005;**289**:L371-L381. DOI: 10.1152/ajplung.00175.2004

[61] Vandenbroucke E et al. Regulation of endothelial junctional permeability. Annals of the new York Academy of Sciences. 2008;**1123**:134-145. DOI: 10.1196/annals.1420.016

[62] Senger DR et al. Tumor cells secrete a vascular permeability factor that promotes accumulation of ascites fluid. Science. 1983;**219**:983-985. DOI: 10.1126/science.6823562

[63] Kimura H et al. Hypoxia response element of the human vascular endothelial growth factor gene mediates transcriptional regulation by nitric oxide: Control of hypoxia-inducible factor-1 activity by nitric oxide. Blood. 2000;**95**:189-197

[64] Gavard J, Gutkind JS. Vegf controls endothelial-cell permeability by promoting the β-arrestin-dependent endocytosis of Ve-Cadherin. Nature

Cell Biology. 2006;**8**:1223. DOI: 10.1038/ncb1486

[65] Rensen S, Doevendans P, Van Eys G. Regulation and characteristics of vascular smooth muscle cell phenotypic diversity. Netherlands Heart Journal. 2007;**15**:100-108. DOI: 10.1007/BF03085963

[66] Garg UC, Hassid A. Nitric oxide-generating vasodilators and 8-bromo-cyclic guanosine monophosphate inhibit mitogenesis and proliferation of cultured rat vascular smooth muscle cells. The Journal of Clinical Investigation. 1989;**83**:1774-1777. DOI: 10.1172/JCI114081

[67] Yu S-M, Hung L-M, Lin C-C. Cgmp-elevating agents suppress proliferation of vascular smooth muscle cells by inhibiting the activation of epidermal growth factor signaling pathway. Circulation. 1997;**95**:1269-1277. DOI: 10.1161/01.CIR.95.5.1269

[68] Cornwell TL et al. Inhibition of smooth muscle cell growth by nitric oxide and activation of camp-dependent protein kinase by Cgmp. American Journal of Physiology-Cell Physiology. 1994;**267**:C1405-C1413. DOI: 10.1152/ajpcell.1994.267.5.C1405

[69] Butt E et al. Camp-and Cgmp-dependent protein kinase phosphorylation sites of the focal adhesion vasodilator-stimulated phosphoprotein (Vasp) in vitro and in intact human platelets. Journal of Biological Chemistry. 1994;**269**:14509-14517

[70] Chen L et al. Vasodilator-stimulated phosphoprotein regulates proliferation and growth inhibition by nitric oxide in vascular smooth muscle cells. Arteriosclerosis, Thrombosis, and Vascular Biology. 2004;**24**:1403-1408. DOI: 10.1161/01.ATV.0000134705.39654.53

[71] Benz PM et al. Differential Vasp phosphorylation controls remodeling of the actin cytoskeleton. Journal of Cell Science. 2009;**122**:3954-3965. DOI: 10.1242/jcs.044537

[72] Tanner FC et al. Nitric oxide modulates expression of cell cycle regulatory proteins: A cytostatic strategy for inhibition of human vascular smooth muscle cell proliferation. Circulation. 2000;**101**:1982-1989. DOI: 10.1161/01.CIR.101.16.1982

[73] Buga GM et al. N G-hydroxy-L-arginine and nitric oxide inhibit Caco-2 tumor cell proliferation by distinct mechanisms. American Journal of Physiology-Regulatory, Integrative and Comparative Physiology. 1998;**275**:R1256-R1264. DOI: 10.1152/ajpregu.1998.275.4.R1256

[74] Takai Y et al. Inhibitory action of guanosine 3′, 5′-monophosphate on thrombin-induced phosphatidylinositol turnover and protein phosphorylation in human platelets. Biochemical and Biophysical Research Communications. 1981;**101**:61-67. DOI: 10.1016/S0006-291X(81)80010-1

[75] Geiger J et al. Role of Cgmp and Cgmp-dependent protein kinase in nitrovasodilator inhibition of agonist-evoked calcium elevation in human platelets. Proceedings of the National Academy of Sciences. 1992;**89**:1031-1035. DOI: 10.1073/pnas.89.3.1031

[76] Furie B, Furie BC. Mechanisms of thrombus formation. New England Journal of Medicine. 2008;**359**:938-949

[77] Frost MC, Reynolds MM, Meyerhoff ME. Polymers incorporating nitric oxide releasing/generating substances for improved biocompatibility of blood-contacting medical devices. Biomaterials. 2005;**26**:1685-1693. DOI: 10.1016/j.biomaterials.2004.06.006

[78] Hopkins SP et al. Achieving long-term biocompatible silicone via covalently immobilized S-nitroso-N-acetylpenicillamine (snap) that exhibits 4 months of sustained nitric oxide release. ACS Applied Materials & Interfaces. 2018;**10**:27316-27325. DOI: 10.1021/acsami.8b08647

[79] Brisbois EJ et al. Reduction in thrombosis and bacterial adhesion with 7 day implantation of S-nitroso-N-acetylpenicillamine (Snap)-doped Elast-eon E2as catheters in sheep. Journal of Materials Chemistry B. 2015;**3**:1639-1645. DOI: 10.1039/C4TB01839G

[80] Shao D et al. Redox modification of cell signaling in the cardiovascular system. Journal of Molecular and Cellular Cardiology. 2012;**52**:550-558. DOI: 10.1016/j.yjmcc.2011.09.009

[81] Bergendi L et al. Chemistry, physiology and pathology of free radicals. Life Sciences. 1999;**65**:1865-1874. DOI: 10.1016/S0024-3205(99)00439-7

[82] Bedard K, Krause K-H. The nox family of ros-generating nadph oxidases: Physiology and pathophysiology. Physiological Reviews. 2007;**87**:245-313. DOI: 10.1152/physrev.00044.2005

[83] Förstermann U, Xia N, Li H. Roles of vascular oxidative stress and nitric oxide in the pathogenesis of atherosclerosis. Circulation Research. 2017;**120**:713-735. DOI: 10.1161/CIRCRESAHA.116.309326

[84] Zimmet JM, Hare JM. Nitroso–redox interactions in the cardiovascular system. Circulation. 2006;**114**:1531-1544. DOI: 10.1161/CIRCULATIONAHA.105.605519

[85] Romero MJ et al. Diabetes-induced coronary vascular dysfunction involves increased arginase activity. Circulation Research. 2008;**102**:95-102. DOI: 10.1161/CIRCRESAHA.107.155028

[86] Xu J et al. Proteasome-dependent degradation of guanosine 5′-triphosphate cyclohydrolase I causes tetrahydrobiopterin deficiency in diabetes mellitus. Circulation. 2007;**116**:944-953. DOI: 10.1161/CIRCULATIONAHA.106.684795

[87] De Montellano PRO, editor. Cytochrome P450: Structure, Mechanism, and Biochemistry. 3rd ed. Boston, MA: Springer Science & Business Media; 2005. DOI: 10.1021/ja041050x

[88] Fleming I. Cytochrome P450 and vascular homeostasis. Circulation Research. 2001;**89**:753-762. DOI: 10.1161/hh2101.099268

[89] De Montellano PRO, De Voss JJ. Substrate oxidation by cytochrome P450 enzymes. In: Cytochrome P450. 3rd ed. Boston, MA: Springer; 2005. pp. 183-245. DOI: 10.1021/ja041050x

[90] Montezano AC, Touyz RM. Molecular mechanisms of hypertension—reactive oxygen species and antioxidants: A basic science update for the clinician. Canadian Journal of Cardiology. 2012;**28**:288-295. DOI: 10.1016/j.cjca.2012.01.017

[91] Veal EA, Day AM, Morgan BA. Hydrogen peroxide sensing and signaling. Molecular Cell. 2007;**26**:1-14. DOI: 10.1016/j.molcel.2007.03.016

[92] Weiss SJ et al. Role of hydrogen peroxide in neutrophil-mediated destruction of cultured endothelial cells. The Journal of Clinical Investigation. 1981;**68**:714-721. DOI: 10.1172/JCI110307

[93] Baas AS, Berk BC. Differential activation of mitogen-activated protein kinases by H_2O_2 and O_2- in vascular smooth muscle cells. Circulation

Research. 1995;**77**:29-36. DOI: 10.1161/01.RES.77.1.29

[94] Cohen G, Heikkila RE. The generation of hydrogen peroxide, superoxide radical, and hydroxyl radical by 6-hydroxydopamine, dialuric acid, and related cytotoxic agents. Journal of Biological Chemistry. 1974;**249**:2447-2452

[95] Phaniendra A, Jestadi DB, Periyasamy L. Free radicals: Properties, sources, targets, and their implication in various diseases. Indian Journal of Clinical Biochemistry. 2015;**30**:11-26. DOI: 10.1007/s12291-014-0446-0

[96] Josephson RA et al. Study of the mechanisms of hydrogen peroxide and hydroxyl free radical-induced cellular injury and calcium overload in cardiac myocytes. Journal of Biological Chemistry. 1991;**266**:2354-2361

[97] Ide T et al. Direct evidence for increased hydroxyl radicals originating from superoxide in the failing myocardium. Circulation Research. 2000;**86**:152-157. DOI: 10.1161/01.RES.86.2.152

[98] Pacher P, Beckman JS, Liaudet L. Nitric oxide and peroxynitrite in health and disease. Physiological Reviews. 2007;**87**:315-424. DOI: 10.1152/physrev.00029.2006

[99] Ago T et al. A redox-dependent pathway for regulating class Ii Hdacs and cardiac hypertrophy. Cell. 2008;**133**:978-993. DOI: 10.1016/j.cell.2008.04.041

[100] Sun J et al. Preconditioning results in S-nitrosylation of proteins involved in regulation of mitochondrial energetics and calcium transport. Circulation Research. 2007;**101**:1155-1163. DOI: 10.1161/CIRCRESAHA.107.155879

[101] He W, Frost MC. Direct measurement of actual levels of nitric oxide (NO) in cell culture conditions using soluble NO donors. Redox Biology. 2016;**9**:1-14. DOI: 10.1016/j.redox.2016.05.002

[102] Li H et al. Nitration and functional loss of voltage-gated K^+ channels in rat coronary microvessels exposed to high glucose. Diabetes. 2004;**53**:2436-2442. DOI: 10.2337/diabetes.53.9.2436

[103] Eiserich JP et al. Microtubule dysfunction by posttranslational nitrotyrosination of A-tubulin: A nitric oxide-dependent mechanism of cellular injury. Proceedings of the National Academy of Sciences. 1999;**96**:6365-6370. DOI: 10.1073/pnas.96.11.6365

[104] Pesse B et al. Peroxynitrite activates Erk via Raf-1 and Mek, independently from Egf receptor and P21ras in H9c2 cardiomyocytes. Journal of Molecular and Cellular Cardiology. 2005;**38**:765-775. DOI: 10.1016/j.yjmcc.2005.02.020

[105] Leeuwenburgh C et al. Reactive nitrogen intermediates promote low density lipoprotein oxidation in human atherosclerotic intima. Journal of Biological Chemistry. 1997;**272**: 1433-1436. DOI: 10.1074/jbc.272.3.1433

[106] Sakakura K et al. Pathophysiology of atherosclerosis plaque progression. Heart, Lung and Circulation. 2013;**22**:399-411. DOI: 10.1016/j.hlc.2013.03.001

[107] Dai G et al. Distinct endothelial phenotypes evoked by arterial waveforms derived from atherosclerosis-susceptible and-resistant regions of human vasculature. Proceedings of the National Academy of Sciences. 2004;**101**:14871-14876. DOI: 10.1073/pnas.0406073101

[108] Ross R. Cell biology of atherosclerosis. Annual Review of Physiology. 1995;**57**:791-804. DOI: 10.1146/annurev.ph.57.030195.004043

[109] Linton MF et al. The role of lipids and lipoproteins in atherosclerosis. In: Endotext [Internet]. 2015. Available from: https://www.ncbi.nlm.nih.gov/books/NBK343489/ [Accessed: 12 November 2018]

[110] Topper JN et al. Identification of vascular endothelial genes differentially responsive to fluid mechanical stimuli: Cyclooxygenase-2, manganese superoxide dismutase, and endothelial cell nitric oxide synthase are selectively up-regulated by steady laminar shear stress. Proceedings of the National Academy of Sciences. 1996;**93**:10417-10422. DOI: 10.1073/pnas.93.19.10417

[111] Gimbrone MA Jr, and G. García-Cardeña: Vascular endothelium, hemodynamics, and the pathobiology of atherosclerosis. Cardiovascular Pathology. 2013;**22**:9-15. DOI: 10.1016/j.carpath.2012.06.006

[112] SenBanerjee S et al. Klf2 is a novel transcriptional regulator of endothelial proinflammatory activation. Journal of Experimental Medicine. 2004;**199**: 1305-1315. DOI: 10.1084/jem.20031132

[113] Halliwell B, Gutteridge JM. Free Radicals in Biology and Medicine. 5th ed. New York, USA: Oxford University Press; 2015. DOI: 10.1093/acprof:oso/9780198717478.001.0001

[114] Rhee SG, Chae HZ, Kim K. Peroxiredoxins: A historical overview and speculative preview of novel mechanisms and emerging concepts in cell signaling. Free Radical Biology and Medicine. 2005;**38**:1543-1552. DOI: 10.1016/j.freeradbiomed.2005.02.026

[115] Rhee SG. Overview on peroxiredoxin. Molecules and Cells. 2016;**39**:1-5. DOI: 10.14348/molcells.2016.2368

[116] Blankenberg S et al. Glutathione peroxidase 1 activity and cardiovascular events in patients with coronary artery disease. New England Journal of Medicine. 2003;**349**:1605-1613. DOI: 10.1056/NEJMoa030535

[117] Mieyal JJ et al. Molecular mechanisms and clinical implications of reversible protein S-glutathionylation. Antioxidants & Redox Signaling. 2008;**10**:1941-1988. DOI: 10.1089/ars.2008.2089

[118] Hare JM. Nitroso-redox balance in the cardiovascular system. New England Journal of Medicine. 2004;**351**: 2112-2114. DOI: 10.1056/NEJMe048269

[119] Flögel U et al. Myoglobin: A scavenger of bioactive NO. Proceedings of the National Academy of Sciences. 2001;**98**:735. DOI: 10.1073/pnas.98.2.735

[120] Herold S, Fago A. Reactions of peroxynitrite with globin proteins and their possible physiological role. Comparative Biochemistry and Physiology Part A: Molecular & Integrative Physiology. 2005;**142**:124-129. DOI: 10.1016/j.cbpb.2005.06.009

[121] Bernhard D, Laufer G. The aging cardiomyocyte: A mini-review. Gerontology. 2008;**54**:24-31. DOI: 10.1159/000113503

[122] Wang JC, Bennett M. Aging and atherosclerosis: Mechanisms, functional consequences, and potential therapeutics for cellular senescence. Circulation Research. 2012;**111**:245-259. DOI: 10.1161/CIRCRESAHA.111.261388

[123] Scheerder ID et al. Biocompatibility of polymer-coated oversized metallic stents implanted in normal porcine coronary arteries. Atherosclerosis. 1995;**114**:105-114. DOI: 10.1016/0021-9150(94)05472-U

[124] Ranganath SH et al. Harnessing the mesenchymal stem cell secretome for the treatment of cardiovascular disease. Cell Stem Cell. 2012;**10**:244-258. DOI: 10.1016/j.stem.2012.02.005

[125] Schmidt-Lucke C et al. Reduced number of circulating endothelial progenitor cells predicts future cardiovascular events: Proof of concept for the clinical importance of endogenous vascular repair. Circulation. 2005;**111**:2981-2987. DOI: 10.1161/CIRCULATIONAHA.104.504340

[126] Smith W. Safety of mechanical thrombectomy and intravenous tissue plasminogen activator in acute ischemic stroke. results of the multi mechanical embolus removal in cerebral ischemia (Merci) trial, Part I. American Journal of Neuroradiology. 2006;**27**:1177-1182

[127] Ridker PM. High-sensitivity C-reactive protein: Potential adjunct for global risk assessment in the primary prevention of cardiovascular disease. Circulation. 2001;**103**:1813-1818. DOI: 10.1161/01.CIR.103.13.1813

Chapter 2

Potassium Channels in the Vascular Diseases

Yan-Rong Zhu, Xiao-Xin Jiang, Peng Ye, Shao-liang Chen and Dai-Min Zhang

Abstract

The vessel wall is an intricate structure composed of three layers: the intima (consisting of endothelial cells), media (consisting of smooth muscle cells and elastic fibers), and externa (consisting of the extracellular matrix scaffold). The homeostasis of the vasculature depends on the consistent function of each layer. In the vascular system, potassium channels are well known to regulate vascular function. The interactions between vascular conditions and membrane potential are complicated. In this chapter, we will focus on the functional regulation of KCa channel, K_{ATP} channel, and K_V channel in the vascular system. Researchers may continuously obtain insights into the functions of these channels and identify new therapeutic targets for vascular diseases.

Keywords: potassium channel, vascular diseases, BK channel, Kv channel, K_{ATP} channel

1. Introduction

The vascular system, which includes an extensive network of arteries, capillaries, and veins, exhibits specific biochemical, cellular, and transport functions. The absorption of essential nutrients and removal of cellular metabolic products both depend on the vasculature [1]. The vessel wall is an intricate structure composed of three layers: the intima, media, and externa. The homeostasis of the vascular system depends on the consistent function of each layer. The thinnest constituent layer is the intima, which consists of a single layer of endothelial cells (ECs) on a basement membrane. Although ECs are typically flat, they are plump or cuboidal in venules composed of numerous endothelial cells [2]. Endothelial cells perform critical functions in all aspects of tissue homeostasis; in addition, they regulate vascular tone by interacting with components of the peripheral nervous system and are related to inflammatory and immunological processes [3, 4]. The media mainly contains smooth muscle cells (SMCs) and elastic fibers. Elastic fibers are mainly a source of structural support, while SMCs play a vital role in maintaining the vascular structure, vascular repair, remodeling, and disease. VSMCs exhibit extraordinary plasticity and undergo remodeling in response to local hemodynamic changes, mechanical forces, hormones, and cytokines [5, 6]. The most remarkable functions of VSMCs are to regulate vascular tone and vessel diameter, which determines blood pressure and tissue perfusion. Some components of vascular contractility, such as the actin cytoskeleton, are required for VSMC proliferation and migration. Circular

RNA, microRNA, and some other transcription factors jointly regulate the expression of smooth muscle α-actin (α-SMA) and the contraction of VSMCs [7, 8]. The outer most layer of the vessel wall is the adventitia. In vertebrates, the adventitia is important because the fibroelastic connective tissue stroma is an important structural component of all tissues. The adventitial stroma contains an extracellular matrix scaffold including fibroblasts, blood and lymphatic vessels, nerve endings, progenitor cells, and immune cells. In one sense, the adventitia is the most complex and heterogeneous compartment of the vessel wall [9].

Ion channels play an important role in the mechanism of action of vasodilators and vasoconstrictors that modulate vascular tone and the effects of disease states, such as hypertension, obesity, and diabetes, which depend on ion channel expression and function. We focus on the basic properties, physiological functions, regulation, and pathological alterations in major classes of K^+ channels that have been detected in VSMCs and/or ECs, including Ca^{2+}-activated K^+ channels, ATP-sensitive K^+ channels, and voltage-gated K^+ channels.

2. Vascular function

In the vascular system, transmembrane voltage regulates vascular function. The interactions between vascular conditions and membrane potential are complicated [10]. The hyperpolarization of the smooth muscle cell membrane potential is evoked by the activation of ion channels, which contributes to vasodilation. A decrease in Ca^{2+} influx resulting from a decrease in the open probability of voltage-dependent calcium channels (Ca_V) and the Ca_V-dependent activation of the sarcoplasmic reticulum are crucial factors contributing to this process [11]. The depolarization of vascular smooth muscle cells causes contraction by opening Ca_V and inducing calcium release. The Ca^{2+}-activated K^+ channels (K_{Ca}) are considered key elements that control vascular tone and blood pressure by modulating membrane hyperpolarization and relaxation. Ca^{2+}-activated K^+ channels, including large conductance Ca^{2+}- and voltage-activated K^+ (BK) channels, intermediate conductance Ca^{2+}-activated K^+ channels (IK), and small-conductance Ca^{2+}-activated K^+ (SK) channels, are widely expressed in the vascular system. Intercellular conduction of electric signals underlies the spread of vasodilation to resistance arteries [10, 12].

3. Ca^{2+}-activated K^+ channel

3.1 The structure of BK channel

BK channels (also known as MaxiK) are widely expressed in vascular smooth muscle cells. Vascular BK channels comprise four pore-forming subunits (BK-α) and four auxiliary subunits: β_1 subunits (BK-β_1) and/or γ_1 subunits (BK-γ_1). BK α, which is encoded by the KCNMA1 gene, has seven transmembrane domains (S0–S6). BK-α has an extra transmembrane segment, S0, and thus its N-terminus is located in the extracellular space. S1–S4 form the voltage-sensor domain (VSD), and S5 and S6 form the ion permeation domain that encompasses the conserved K^+ filter (TVGYG) in the pore loop. The C-terminus of BK channels modulates the voltage sensor and affects the pore, thus influencing channel opening. The ability of specific BK channels to open as a function of Ca^{2+} concentration or as a function of voltage sensors is due to the use of alternative splice sites [13–15]. The C-terminus contains two homologous structural units termed "regulators of conductance for K^+": the proximal portion RCK domain (RCK1) and the distal portion RCK domain (RCK2).

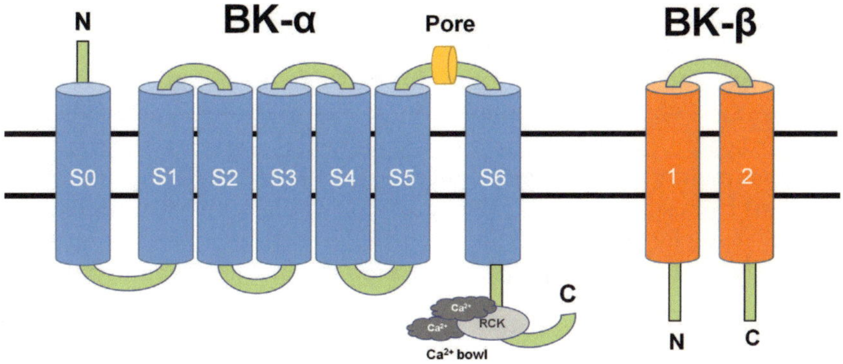

Figure 1.
Schematic structure of one BKα-subunit consisting of 7 transmembrane domains (S0–S6). S1–S4 constitute the voltage-sensing unit, and the S5-P loop-S6 form the ion permeation domain. The Ca^{2+} bowl is a high-affinity divalent cation-binding domain, and the RCK domain (regulator of conductance for K^+) in the C-terminal region is responsible for Ca^{2+} sensitivity. The presence of the β1 subunit increases Ca^{2+} sensitivity and thus channels activity.

The RCK1 domain is related to the formation of the Ca^{2+}-binding site, and the RCK2 contains a "high-affinity" Ca^{2+} bowl domain. The "gating ring" is a Ca^{2+}-sensing apparatus composed of four pairs of RCK1 and RCK2 domains, and the function of gating ring is responsible for allosteric activation of BK channel by Ca^{2+} binding [15–19] (**Figure 1**). Four types of β subunits (BK-β) and four types of γ subunits (BK-γ) modulate almost all aspects of the pharmacological actions and physiological processes mediated by BK channels. The functional mechanism of BK channels regulated by β and γ auxiliary subunits is extremely complicated but is crucial to our understanding of its implications in vascular diseases. In the vasculature, BK-β_1, which is encoded by KCNMB1, is the dominant isoform in VSMCs, and the dysfunction of the β_1 is associated with diabetes, hypertension, and other vascular diseases. Knockout of the BK-β_1 gene produces a remarkable decrease in the Ca^{2+} sensitivity of the channel. In addition, coexpression of the BK-β_1 subunit with the BK-α subunit dramatically alters the calcium sensitivity, similar to the results observed in native VSMCs [15, 20, 21]. As an auxiliary subunit of the BK channels, BK-γ also affects BK channel activity by modulating the voltage and Ca^{2+} sensitivity. The BK-γ subunit has the ability to regulate vascular tone, and knockdown of BK-γ subunits contributes to pressure-induced vasoconstriction and a decrease in the activity of functional BK channels [15, 22, 23].

3.2 The physiological function of BK channel

Hundreds of proteins (such as β-catenin and caveolins) are reported to interact with BK channels in various systems *in vitro* and/or *in vivo*. The mutual effect between BK channel and these proteins regulates the BK channel functions and influences the biological pathways mediated by BK channels [24–26]. However, a key characteristic of BK channels is their ability to couple with calcium channels that mediate the increase of intracellular Ca^{2+}. BK channels can prevent Ca^{2+} channels from further activation and limit Ca^{2+} influx. In smooth muscle cells, ryanodine receptors cause local, transient calcium release events from the endoplasmic reticulum. These spontaneous calcium release events lead to the activation of nearby BK channels, which induce membrane hyperpolarization. This kind of potassium current is called the spontaneous transient outward current (STOC), and by blocking, STOC contributes to the increased vascular muscle tone [24, 27, 28]. Accordingly, BK channel is a key regulator to induce vasodilatation.

3.3 The function of BK channel in diabetes and hypertension

Diabetes is an independent risk factor for vascular diseases and is associated with increased risks of vascular complications, such as coronary artery disease, stroke, nephropathy, neuropathy, and retinopathy [29]. Vascular BK channel dysfunction is mainly due to a significant downregulation of BK-β_1 subunit expression in vessels from subjects with T1DM and T2DM. The activity of BK channels is regulated by many factors, such as angiotensin II, reactive oxygen species (ROS), nitric oxide (NO), carbon monoxide (CO), and protein kinase A- and protein kinase C-mediated signaling pathways.

According to Lu et al., the ROS signaling cascade facilitates Forkhead box O subfamily transcription factor-3a (FOXO-3a)-dependent F-box-only protein (FBXO)-mediated BK-β_1 degradation and leads to the dysfunction of diabetic BK channels. In diabetic mouse aortas and in high glucose-cultured human coronary arterial smooth muscle cells, p-Akt (S473) levels are decreased, and the level of protein kinase C (PKC) β, which stimulates ROS generation and contributes to diabetic cardiovascular complications in diabetic rats, is distinctly increased [29, 30]. This group also revealed that the nuclear factor erythroid-2-related factor 2 (Nrf2) signaling pathway plays a significant role in regulating coronary BK channel function and vasodilation in mice with high-fat diet (HFD)-induced obesity/diabetes [31].

Hypertension, which is characterized by increased arterial tone, is another risk factor for cardiovascular diseases. Substantial evidence shows decreased expression of the BK-β_1 subunit that is considered to contribute to the development of vascular dysfunction during hypertension. Loss-of-function mutations in BK-β_1 decrease the prevalence of diastolic hypertension in humans [32]. Recently, the regulated trafficking of BK channel subunits (including α subunit and auxiliary β_1 and γ subunits) has been accepted as a functional mechanism to modulate arterial contractility. Endothelin-1 (ET-1) is a vasoconstrictor that activates protein kinase C (PKC) and stimulates PKC-mediated phosphorylation of Rab11A at serine 177. Subsequently, surface β_1 trafficking is reduced, resulting in a decrease in BK channel currents and vasoconstriction [33, 34].

3.4 BK channel in ECs

BK channels are expressed in both VSMCs and endothelial cells [35, 36]. In the majority of the systemic vasculature, endothelial BK channels are electrically quiescent, but may be disinhibited under pathophysiological conditions [37]. Hydrogen sulfide (H_2S) is an important, endogenously generated gaseous signaling molecule. H_2S-mediated vasodilation involves the activation of endothelial BK channels, which depends on Ca^{2+} influx through endothelial transient receptor potential vanilloid-4 (TRPV4) channels [38].

Using the whole-cell recording technique, Dong et al. examined the effect of CO on the activity of BK channels. The application of exogenous CO-activated BK channels in endothelial cells and the stimulation of endogenous CO production increased BK channel activity in human umbilical vein endothelial cells (HUVECs). Stimulation of soluble guanylate cyclase (sGC) production is responsible for the early stage, but not the latter stage, of this process. The CO-induced activation of BK channels plays an essential role in modulating vascular function. In endothelial cells, BK channels are activated by CO and induce the hyperpolarization of the membrane potential. Afterwards, the driving force for Ca^{2+} influx increases, and the increase in the intracellular Ca^{2+} concentration stimulates NO generation, which diffuses into the smooth muscle cells to activate BK channels [35].

Another key factor that interacts with BK channels and likely exerts a negative regulatory effect on channel activity is caveolin-1 (Cav-1). Under normal conditions, Cav-1 limits the contribution of the BK channels to EDHF-mediated arteriolar dilation. In obesity, the decreased expression of Cav-1 increases the contribution of the BK channels to EDHF-mediated arteriolar dilation, which seems essential for maintaining vascular homeostasis [39]. Chronic hypoxia (CH) enhances the activity of BK channels in ECs and alters vasoreactivity via the loss of an inhibitory effect of Cav-1. Under this condition, BK channels in ECs display a similar unitary conductance but greater Ca^{2+} sensitivity than BK channels from vascular smooth muscle cells [40].

Anandamide is an endogenous ligand for specific G-protein-coupled cannabinoid type 1 (CB_1) and type 2 (CB_2) receptors. In the cardiovascular system, anandamide acts as a direct BK_{Ca} opener, and vasodilatory responses to cannabinoids are thought to require a G-protein-coupled receptor (GPCR) located on endothelial cells, the activation of which results in the direct modification of BK_{Ca} channel activity and BK_{Ca}-dependent vasodilation. BK_{Ca} channels act as cellular sensors for cannabinoids in *in vitro* and *in situ* endothelial cells [40]. The mechanism of action of anandamide on endothelial cells was not previously believed to require CB1, CB_2, or non-CB_1/CB_2 receptors, but was related to direct modulation of the BK_{Ca} channel gating without modification of unitary conductance [41]. However, the roles of BK_{ca} in endothelial cells observed in response to *in vitro* and *in situ* cannabinoid-induced vasodilation are undisputed.

3.5 Structures of SK and IK channel

SK and IK channels are two distinct types of voltage-independent K_{Ca} channels; these channels exhibit a close association between their calcium sensitivity and calmodulin [42]. In contrast to intestinal smooth muscle, little evidence is available suggesting a functional role for SK channels in vascular smooth muscle cells, although an unidentified apamin (a specific blocker of SK channel channels)-sensitive and voltage-dependent conductance has been reported [43]. In healthy and freshly isolated vascular smooth muscle cells, IK channels are expressed at very low levels. In contrast, the expression of IK channels increases when the vascular system is impaired, and this phenomenon also appears in proliferating smooth muscle cells [44].

The family of SK channels consists of three members: SK1 (also known as KCa2.1), which is encoded by the KCNN1 gene; SK2 (also known as KCa2.2), which is encoded by the KCNN2 gene; and SK3 (also known as KCa2.3), which is encoded by the KCNN3 gene. SK channels consist of six transmembrane regions (TMs) and a single pore loop, with four subunits located around a central pore. Both the N-terminus and C-terminus are oriented toward the cytoplasm. SK channels have no charged amino acids in the fourth TM domain, which is usually an important component of a voltage sensor. SK channels are activated and deactivated solely as a consequence of Ca^{2+} binding or release [45]. SK channels are heteromeric complexes that comprise pore-forming α subunits and the Ca^{2+}-binding protein calmodulin (CaM) (**Figure 2**). CaM is not only necessary for Ca^{2+} sensitivity but also critical for the trafficking of SK channels. CaM binds to and activates its target proteins in both Ca^{2+}-replete and Ca^{2+}-depleted forms. CaM mutants affect the interaction of CaM with its target proteins [45, 46]. CaM binds to a highly conserved CaM-binding domain (CaMBD) residing within the C-terminus of the SK channels that is located immediately distal to the sixth transmembrane segment [47, 48]. Maria A. Schumacher et al. explored the structure of the CaMBD/Ca^{2+}/CaM complex, and in this complex, CaM binds three α-helices instead of one, and the N-lobe and C-lobe of each CaM molecule contact different CaMBD monomers. The structure of the CaMBD/Ca^{2+}/CaM complex provides detailed information about both Ca^{2+}-dependent and Ca^{2+}-independent CaM interactions in a single complex [48].

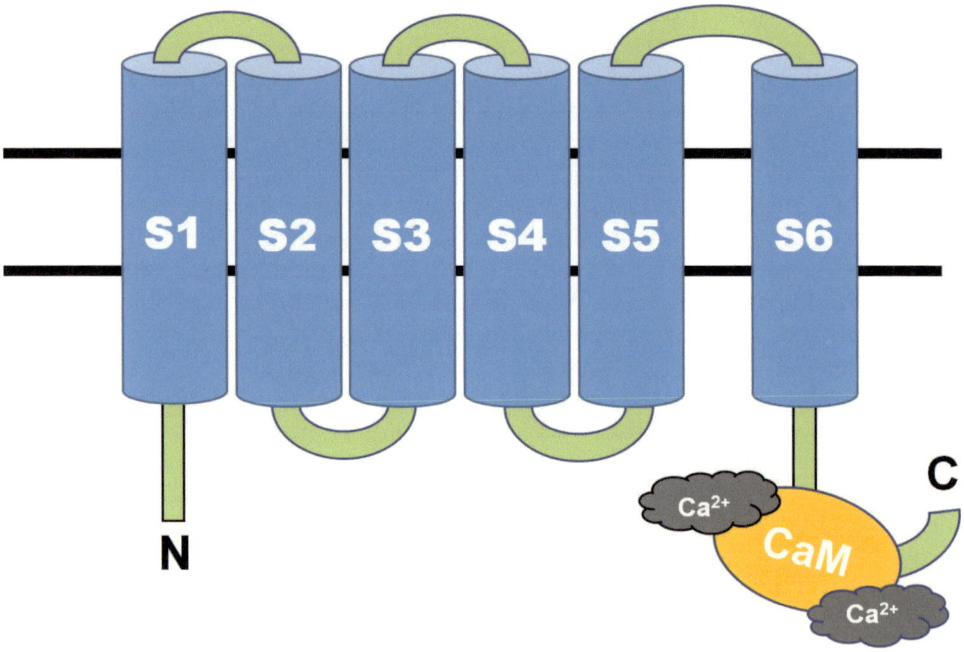

Figure 2.
Schematic of IK and three subtypes of SK (SK1, SK2, and SK3). SK3 and IK are thought to be the predominant KCa channels expressed in systemic vascular endothelia. The basic structure consisting of six transmembrane domains (S1–S6). The constitutively bound calmodulin (CaM), at the C-terminus. The founding member of Ca^{2+} binding proteins is CaM, a small, acidic, modular protein endowed with gymnastic-like flexibility that chelate Ca^{2+} ions.

IK channel (also known as KCa 3.1) is widely expressed in cells of the immune system and red and white blood cells, where it plays an important role in cellular activation, migration, and cytokine production [49–51]. Moreover, KCa3.1 is also expressed in dedifferentiated vascular smooth muscle cells, fibroblasts, and the vascular endothelium, where the channel is involved in the EDH response [52, 53]. The KCa3.1 channel is a tetrameric membrane protein with each subunit (comprising 427 amino acids) organized in six transmembrane segments, S1–S6, with a pore motif between segments 5 (S5) and 6 (S6). The channel assembly and trafficking are regulated by the constitutively bound calmodulin (CaM) molecule, which also confers Ca^{2+} sensitivity [54, 55]. Ca^{2+} binds to the CaM-KCa3.1 complex in the C-terminus of KCa3.1. As the CaM-binding domain of KCa3.1 is directly connected to the S6 transmembrane helix, activation of the channel gate at the level of the selectivity filter might depend upon the coupling between each of the channel pore helices and the associated S6 transmembrane segment. The interactions of the KCa3.1 pore helix with the S5 and S6 transmembrane segments also contribute to setting POmax, which is one of the distinguishing features of the Ca^{2+}-dependence of the KCa3.1 channel [55].

3.6 SK and IK channel and EDH responses

In small arterial and arteriolar ECs, KCa channels are activated by intrinsic spontaneous or receptor-mediated Ca^{2+} events, which contribute to the hyperpolarization of SMCs and vasodilation through a NO-independent process. This response is known as endothelium-dependent hyperpolarization (EDH), and it is the predominant mechanism in ECs [56]. SK and IK channels operate in parallel to generate EDH; they contribute to smooth muscle hyperpolarization and vasorelaxation, and the hyperpolarization of the endothelial cells in turn increases calcium influx by increasing the driving force for this ion, but the channels can be activated independently. SK channels are distributed throughout the endothelial cell membrane, but cluster in the proximity of the large gap junctions between endothelial cells. In contrast, IK channels are only present in detectable amounts at endothelial cell projections toward adjacent smooth muscle cells, where they can form myoendothelial gap junctions [57]. EDHF-mediated responses play a physiological role in regulating vascular resistance. In rats, the hypotensive response to endothelium-dependent agonists, such as acetylcholine and bradykinin, is rapidly compensated within 1 day after treatment with the NOS inhibitor N^{ω}-nitro-L-arginine methyl ester (L-NAME). The compensatory relaxation is mediated by the activation of SK and IK channels. Endothelial dysfunction, measured as a reduced endothelium-dependent hypotensive response, does not develop after the inhibition of NOS activity [58].

Over the past two decades, studies examining the physiological role of hydrogen sulfide (H_2S) have received increasing attention. Cystathionine γ-lyase (CSE) generates H_2S under physiological conditions, and a CSE deletion in mice reduces H_2S levels in some tissues, including the aorta. These mice lacking the CSE gene display pronounced hypertension, indicating that H_2S is a physiological vasodilator and regulator of blood pressure [59]. In many ways, either H_2S itself is an EDHF or H_2S releases EDHF from the endothelium [60, 61]. The resting membrane potential of SMCs is increased in CSE knockout mice, and methacholine (a cholinergic-muscarinic agonist)-induced endothelium-dependent relaxation of mesenteric arteries was abolished. Methacholine hyperpolarizes SMCs in endothelium-intact mesenteric arteries from wild-type mice. The application of atropine (a muscarinic antagonist) or charybdotoxin and apamin, which block SK/IK channels, or knockout of the CSE gene in mice inhibited this effect. Simultaneously, the expression of SK2.3, but not the IK3.1 channel, in vascular tissues was increased by H_2S and decreased by a CSE inhibitor or CSE gene knockout [51]. Moreover, insufficient H_2S levels impair EDHF-induced vascular relaxation by increasing oxidative stress and IK inactivation in mice with type 2 diabetes mellitus (T2DM)/hyperhomocysteinemia (HHcy) [62].

The activation of SK/IK channels may regulate electrical conduction along the endothelium of intact vessels, and some factors limit this process, such as myoendothelial coupling to SMCs, perivascular nerve activity, and circulating vasoactive agents. Using intact EC tubes produced after the dissociation of SMCs with mild enzymatic digestion, Behringer and his colleagues found that activation of SK/IK channels impairs the transmission between axial signals. This effect results from a decrease in membrane resistance (r_m) that dissipates charge as current flows from cell to cell along the endothelium [63]. Another group verified these results and further assessed impairments in electric conduction along the endothelium of resistance arteries through the enhanced activation of SK/IK channels. Fresh EC tubes were isolated from resistance arteries in skeletal muscle from different groups of mice. Group 1 included young mice (approximately 4–6 month old), group 2 included middle-aged mice (approximately 12–14 month old), and group 3 included old mice (approximately 24–46 month old). The ability of the endothelium of skeletal muscle resistance arteries to conduct electric signals is impaired with aging. The dual function of SK/IK channels in initiating and modulating electric signaling along the endothelium is altered with aging. By increasing the activation of SK/IK channels (particularly the IK channel), aging promotes hyperpolarization of the endothelium while decreasing its ability to conduct electrical signals. Oxidative stress activates SK/IK channels in the resistance artery endothelium via the action of hydrogen peroxide (H_2O_2) [64].

4. ATP-sensitive potassium channel in the vascular system

Functional K_{ATP} channels are hetero-octameric membrane protein complexes that comprise four inward-rectifier potassium channel 6 (Kir6, either Kir6.1 or Kir6.2) subunits and four ABCC (ATP-binding cassette, subfamily C) family member sulfonylurea receptor (SUR) subunits, including SUR1, SUR2A, or SUR2B. The Kir6 subunit (Kir6.1 or Kir6.2) has two membrane-spanning regions (M1 and M2) with intracellular N- and C-termini. The latter two are alternative splice variants, differing from each other only in the C-terminal 42 amino acids. The SURx subunit has 17 transmembrane regions, arranged in three domains: TMD0, TMD1, and TMD2. A conserved intracellular nucleotide binding fold (NBF1), with Walker A and Walker B domains, exists between TMD1 and TMD2. A second intracellular nucleotide binding fold (NBF2) exists in the C-terminus region of the protein. It is thought that NBF1 binds (and hydrolyzes) MgATP, whereas MgADP binds primarily to NBF2 to stimulate channel activity (**Figure 3**). KATP channels are expressed in a variety of cell types, including cardiac, smooth, and skeletal muscles, with tissue-specific diversity in the receptor subtypes. While pancreatic K_{ATP} channels are associated with SUR1, cardiovascular channels interact with SUR2 subtypes. In VSMC, SUR2B interacts with Kir6.1 to form K_{ATP}, and more rarely, Kir6.2 may be the ion pore-forming subunit. The Kir6 channel pore-forming subunits are the ATP sensor, and their activity is regulated by PIP2. K_{ATP} channels are inhibited by elevated intracellular ATP and stimulated by ADP under physiological conditions [46, 65–67].

In blood vessels, K_{ATP} channels remain closed under normal physiological conditions; however, they are activated when the cell metabolism is disturbed by hypoxia or ischemia, resulting in an efflux of potassium ions and membrane hyperpolarization. The decreased membrane excitability leads to a shortened cardiac action potential, inhibition of neurotransmitter release, and relaxation of vascular smooth muscles, which play key roles in limiting cellular damage or regulating blood pressure [68, 69]. In skeletal muscle arteries and arteries, alterations in metabolic activity induce changes in local oxygen tension and are an important mediator of

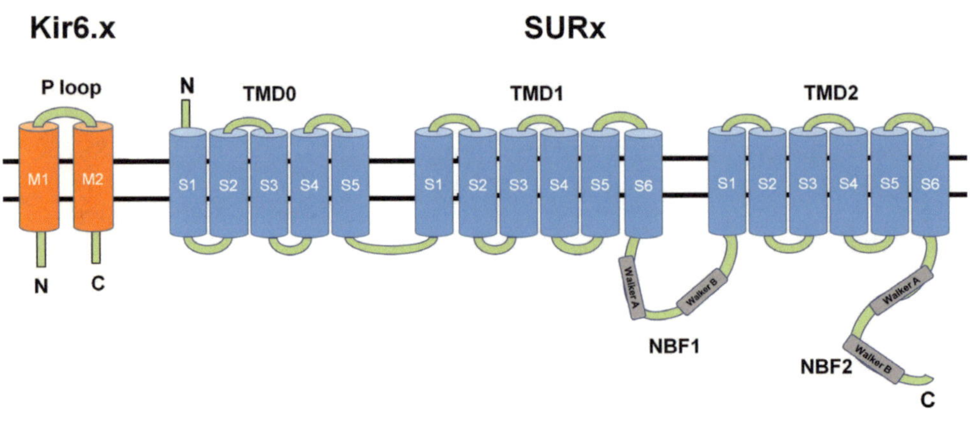

Figure 3.
K_{ATP} channels are hetero-octameric membrane protein complexes that are composed of four inward-rectifier potassium channel 6 (Kir6.x) subunits and sulfonylurea receptor (SURx) subunits. The Kir6.x subunit (Kir6.1 or Kir6.2) has two membrane-spanning regions (M1 and M2) with intracellular N- and C-termini. Two SURx subunits have been described: SUR1 and SUR2 (SUR2A or SUR2B). The latter two are alternative splice variants, differing from each other only in the C-terminal 42 amino acids. The SURx subunit has 17 transmembrane regions, arranged in three domains: TMD0, TMD1, and TMD2. A conserved intracellular nucleotide binding fold (NBF1), with Walker A and Walker B domains, exists between TMD1 and TMD2. A second intracellular nucleotide binding fold (NBF2) exists in the C-terminus region of the protein. It is thought that NBF1 binds (and hydrolyzes) MgATP, whereas MgADP binds primarily to NBF2 to stimulate channel activity.

vasomotor responses. Vasodilation (hypoxic vasodilation) is caused by decreased oxygen tension, and vasoconstriction (hyperoxic vasoconstriction) is caused by the increased oxygen tension [70, 71]. K_{ATP} channels are known to link cell metabolism and cell membrane potential, and decreased oxygen tension results in a depletion of intracellular ATP levels, which contributes to the opening of K_{ATP} channels and the subsequent hyperpolarization and relaxation of the VSMCs [71].

Renal hyperfiltration is a main characteristic of the early stage of type 1 diabetes mellitus (DM), and altered renal hemodynamics promote the eventual development of diabetic nephropathy. The hyperfiltration state is ascribed to the dilation of afferent arterioles and diminished responsiveness of this vascular segment to various vasoconstrictors, while the diameters of efferent arterioles and vasoconstrictor responsiveness are typically unaltered [72–74]. The membrane potential (E_m) and afferent arteriolar dilation are closely related in subjects with DM. K_{ATP} channels are quiescent in normal rats but exert a vasodilatory effect on afferent arteriolar tone during the hyperfiltration stage of diabetes. Increases in both the functional availability and basal activation of K_{ATP} channels promote afferent arteriolar vasodilation during the early stage of DM, changes that likely contribute to the etiology of diabetic hyperfiltration [73]. However, the involvement of K_{ATP} channels in the renal afferent arteriolar dilation during the early stage of DM is still controversial. Additional studies are needed to completely elucidate the potential roles of renal vascular K_{ATP} channels in early diabetic hyperfiltration [74].

5. K_V channel in VSMCs

K_V channels comprise a large family of channels that are expressed in both excitable and nonexcitable cells. In excitable cells, such as neurons or cardiac myocytes, the control of the resting membrane potential (resting E_m) and frequency and duration of action potentials depend on K_V channels. In nonexcitable tissues, these channels are involved in various processes ranging from secretion to cell proliferation [75]. In humans, K_V channels are encoded by 40 genes, and each Kv channel gene encodes a single protein; functional Kv channels are divided into 12 subfamilies (K_V1–K_V12). All mammalian K_V channels consist of four α-subunits and six transmembrane α-helical segments (S1–S6), and a membrane-reentering P-loop forms each α-subunit. This ion conduction pore is lined by four S5–P–S6 sequences. The four S1–S4 segments, each containing four positively charged arginine residues in the S4 helix, act as voltage sensor domains and "gate" the pore by "pulling" on the S4–S5 linker [76, 77]. The large number of K_V channel genes combined with the possibility of heterotetramerization creates a large functional diversity of K_V currents. This diversity is increased by the interactions of these channels with accessory proteins that are capable of modulating the gating properties and assist in trafficking and multimerization [75]. Since the K_V channel subunits form homo and heterotetramers, the biophysical properties, physiological regulatory mechanisms, and pharmacological properties of these channels vary. Although the K_V1.1–1.6 mRNAs have been detected in rat cerebral arteries, only the K_V1.2 and 1.5 proteins were detected, suggesting that in the cerebral vasculature, the functional K_V channel is a K_V1.2/1.5 heterotetramer. Members of the K_V1 and K_V2 family are postulated to be the predominant Kv channels that regulate arterial tone (**Table 1**) [78, 79].

K_V channels regulate membrane potential. Numerous studies have been conducted to explore the mechanisms by which these channels affect vascular tone in subjects with hypertension. Under Ca^{2+}-replete conditions, K_V currents in arterial SMCs from hypertensive animals are altered. K_V1.2 is expressed at higher levels, whereas K_V1.5 is expressed at the same levels in SMCs from hypertensive animals

Family	Subtype in vascular	Gene name	Inhibitor
Ca^{2+}-activated K^+ channels (Kca)	KCa1(BKCa)	KCNMA1 KCNMB1-4	Iberiotoxin (IBTX) Charybdotoxin Paxilline
	KCa2(SKCa)	KCNN1-3	Apamin UCL1684 TRAM-34 Psora-4
	KCa3(IKCa)	KCNN4	Charybdotoxin Clotrimazole TRAM-34 NS6180 Psora-4
ATP-sensitive K^+ channels (K_{ATP})	Kir6.1	KCNJ8	Glibenclamide Tolbutamide
	Kir6.2	KCNJ11	Tolbutamide Glibenclamide ML133
Voltage-gated K^+ channels (K_V)	Kv1	KCNA	4-Aminopyridine(4-AP) Tetraethylammonium (TEA) Correolide α-Dendrotoxin
	Kv2	KCNB	4-Aminopyridine(4-AP) Tetraethylammonium (TEA) Ba^{2+} SsmTx-1
	Kv7	KCNQ	TEA Linopirdine XE991 Chromanol 293B

Table 1.
The three family members of K^+ channels.

than in cells from normal animals [80]. Li et al. confirmed the effect of exercise training on alterations in K_V expression in thoracic aorta smooth muscle cells from spontaneously hypertensive rats (SHR). Rats were divided into three groups, a sedentary spontaneously hypertensive group (SHR-SED) and an exercise training spontaneously hypertensive group (SHR-EX), along with age-matched Wistar-Kyoto rats (WKYs) as the control group. Significantly, lower levels of the $K_V1.2$ and $K_V1.5$ channels were detected in the SHR-SED group than in the WKY group, while this decrease was inhibited in the SHR-EX group. Exercise training reverses the pathological expression of the $K_V1.2$ and $K_V1.5$ channels in aortic myocytes from SHRs, and thus is one of the favorable effects of exercise training on large conduit arteries [81].

The $K_V1.5$ protein is present in the vascular smooth muscle layer of both porcine and human coronary arteries, including microvessels [82]. The mean arterial pressure (MAP), myocardial blood flow (MBF), and ejection fraction (EF) have been measured in wild-type (WT) mice, mice null for $K_V1.5$ channels ($K_V1.5^{-/-}$), and mice with inducible, smooth muscle-specific expression of $K_V1.5$ channels (on $K_V1.5^{-/-}$ and wild type backgrounds). During a norepinephrine (NE) infusion, significantly lower values for EF and MSF were observed in $K_V1.5^{-/-}$ mice than in WT mice. The expression of $K_V1.5$ channels in smooth muscle in mice on the null background rescued this phenotype of impaired metabolic dilation, indicating that Kv1.5 channels in vascular smooth muscle play a critical role in coupling myocardial blood flow to cardiac metabolism. The absence of these channels disassociates metabolism from flow, resulting in cardiac pump dysfunction and tissue hypoxia [83].

In addition to the K_V1 family, the K_V7 ($K_V7.4$ and $K_V7.5$) family has recently been shown to be a major determinant of vascular tone. K_V7 is expressed at similar levels in the murine aorta, carotid, femoral, and mesenteric artery, whereas the expression of $K_V7.4$ and $K_V7.5$ is greater than or equal to $K_V7.1$ [84]. By activating $K_V7.4$ channels, the application of 4-aminopyridine (4-AP) to noradrenaline-preconstricted rat mesenteric arteries contributes to the relaxation of the vessel [85]. The interaction between microRNAs (miRs) and $K_V7.4$ is also important in the vasculature. The expression of miR153 is increased in mesenteric, renal, and thoracic aortic arteries from SHRs compared to NT rats. In SHRs, the expression of $K_V7.4$ is decreased, whereas this change is not consistently associated with a change in transcript level because a difference in mRNA levels was not observed in renal and mesenteric arteries between SHRs and normotensive (NT) rats. In a study using synthetic RNA molecules, miR153 repressed the translation of $K_V7.4$ mRNA rather than degrading the transcript. Thus, miRs regulate the expression of Kv7.4 in the vasculature, and this post-transcriptional regulatory pathway might contribute to vascular dysfunction [86].

6. Conclusions and further perspective

Studies performed over several decades have substantially improved our knowledge of the expression of K^+ channels in the vascular system and their roles in regulating vascular tone and tissue perfusion. Dysfunctional K^+ channels can alter vascular homeostasis through heterogeneous and complex mechanisms. K^+ channels are targets for gene therapy for hypertension. The BK β_1 subunit, K_V 1.5, K_V 7.4, and some other genes should be studied as gene therapy targets. However, some remaining questions still deserve to study. How these K^+ channels work in microvasculature? How can we design better drugs to target these channels with some degree of specificity?

Acknowledgements/funding

This study was supported by the National Natural Science Foundation of China (81370304); Natural Science Foundation of Jiangsu Province (BK20151085); Jiangsu Provincial Key Research and Development Program (BE2018611); the 10th Summit of Six Top Talents of Jiangsu Province (2016-WSN-185); Medical Science and technology development Foundation of Nanjing Department of Health (YKK15101, ZKX16048).

Conflicts of interest

The authors have no conflict of interest to declare.

Author details

Yan-Rong Zhu†, Xiao-Xin Jiang†, Peng Ye, Shao-liang Chen and Dai-Min Zhang*
Department of Cardiology, Nanjing First Hospital, Nanjing Medical University, Nanjing, Jiangsu, China

*Address all correspondence to: daiminzh@126.com

† These authors contributed equally.

References

[1] Pugsley MK, Tabrizchi R. The vascular system: An overview of structure and function. Journal of Pharmacological and Toxicological Methods. 2000;**44**(2):333-340. DOI: 10.1016/s1056-8719(00)00125-8

[2] Girard JP, Springer TA. High endothelial venules (HEVs): Specialized endothelium for lymphocyte migration. Immunology Today. 1995;**16**(9):449-457. DOI: 10.1016/0167-5699(95)80023-9

[3] Maoz BM, Herland A, Fitzgerald EA, et al. A linked organ-on-chip model of the human neurovascular unit reveals the metabolic coupling of endothelial and neuronal cells. Nature Biotechnology. 2018;**36**(9):865-874. DOI: 10.1038/nbt.4226

[4] Špiranec K, Chen W, Werner F, et al. Endothelial C-type natriuretic peptide acts on pericytes to regulate microcirculatory flow and blood pressure. Circulation. 2018;**138**(5):494-508. DOI: 10.1161/CIRCULATIONAHA.117.033383

[5] Grootaert M, Moulis M, Roth L, et al. Vascular smooth muscle cell death, autophagy and senescence in atherosclerosis. Cardiovascular Research. 2018;**114**(4):622-634. DOI: 10.1093/cvr/cvy007

[6] Bakker EN, Pistea A, Spaan JA, et al. Flow-dependent remodeling of small arteries in mice deficient for tissue-type transglutaminase: Possible compensation by macrophage-derived factor XIII. Circulation Research. 2006;**99**(1):86-92. DOI: 10.1161/01.RES.0000229657.83816.a7

[7] Yu K, Zheng B, Han M, et al. ATRA activates and PDGF-BB represses the SM22alpha promoter through KLF4 binding to, or dissociating from, its cis-DNA elements. Cardiovascular Research. 2011;**90**(3):464-474. DOI: 10.1093/cvr/cvr017

[8] Sun Y, Yang Z, Zheng B, et al. A novel regulatory mechanism of smooth muscle α-actin expression by NRG-1/circACTA2/miR-548f-5p axis novelty and significance. Circulation Research. 2017;**121**(6):628-635. DOI: 10.1161/CIRCRESAHA.117.311441

[9] Stenmark KR, Yeager ME, El KK, et al. The adventitia: Essential regulator of vascular wall structure and function. Annual Review of Physiology. 2013;**75**:23-47. DOI: 10.1146/annurev-physiol-030212-183802

[10] Puro DG, Kohmoto R, Fujita Y, et al. Bioelectric impact of pathological angiogenesis on vascular function. Proceedings of the National Academy of Sciences of the United States of America. 2016;**113**(35):9934-9939. DOI: 10.1073/pnas.1604757113

[11] Nelson MT, Patlak JB, Worley JF, et al. Calcium channels, potassium channels, and voltage dependence of arterial smooth muscle tone. The American Journal of Physiology. 1990;**259**(1 Pt 1):C3-C18. DOI: 10.1152/ajpcell. 1990.259.1.C3

[12] Schmid J, Muller B, Heppeler D, et al. The unexpected role of calcium-activated potassium channels: Limitation of NO-induced arterial relaxation. Journal of the American Heart Association. 2018;**7**(7):e007808. DOI: 10.1161/JAHA.117.007808

[13] Meera P, Wallner M, Song M, et al. Large conductance voltage- and calcium-dependent K^+ channel, a distinct member of voltage-dependent ion channels with seven N-terminal transmembrane segments (S0–S6), an extracellular N terminus, and an intracellular (S9–S10) C terminus. Proceedings of the National Academy

of Sciences of the United States of America. 1997;**94**(25):14066-14071

[14] Saito M, Nelson C, Salkoff L, et al. A cysteine-rich domain defined by a novel exon in a slo variant in rat adrenal chromaffin cells and PC12 cells. The Journal of Biological Chemistry. 1997;**272**(18):11710-11717

[15] Zhu Y, Ye P, Chen S, et al. Functional regulation of large conductance Ca^{2+}-activated K^{+} channels in vascular diseases. Metabolism. 2018;**83**:75-80. DOI: 10.1016/j.metabol.2018.01.008

[16] Yuan P, Leonetti MD, Hsiung Y, et al. Open structure of the Ca^{2+} gating ring in the high-conductance Ca^{2+}-activated K^{+} channel. Nature. 2011;**481**(7379):94-97. DOI: 10.1038/nature10670

[17] Zhang G, Huang SY, Yang J, et al. Ion sensing in the RCK1 domain of BK channels. Proceedings of the National Academy of Sciences of the United States of America. 2010;**107**(43):18700-18705. DOI: 10.1073/pnas.1010124107

[18] Yusifov T, Savalli N, Gandhi CS, et al. The RCK2 domain of the human BKCa channel is a calcium sensor. Proceedings of the National Academy of Sciences. 2008;**105**(1):376-381. DOI: 10.1073/pnas.0705261105

[19] Yang J, Krishnamoorthy G, Saxena A, et al. An epilepsy/dyskinesia-associated mutation enhances BK channel activation by potentiating Ca^{2+} sensing. Neuron. 2010;**66**(6):871-883. DOI: 10.1016/j.neuron.2010.05.009

[20] Lu T, Chai Q, Jiao G, et al. Downregulation of BK channel function and protein expression in coronary arteriolar smooth muscle cells of type 2 diabetic patients. Cardiovascular Research. 2018. DOI: 10.1093/cvr/cvy137 [Epub ahead of print]

[21] Pluger S, Faulhaber J, Furstenau M, et al. Mice with disrupted BK channel beta1 subunit gene feature abnormal Ca(2+) spark/STOC coupling and elevated blood pressure. Circulation Research. 2000;**87**(11):E53-E60

[22] Yan J, Aldrich RW. LRRC26 auxiliary protein allows BK channel activation at resting voltage without calcium. Nature. 2010;**466**(7305):513-516. DOI: 10.1038/nature09162

[23] Evanson KW, Bannister JP, Leo MD, et al. LRRC26 is a functional BK channel auxiliary gamma subunit in arterial smooth muscle cells. Circulation Research. 2014;**115**(4):423-431. DOI: 10.1161/CIRCRESAHA.11 5.303407

[24] Kim H, Oh KH. Protein network interacting with BK channels. International Review of Neurobiology. 2016;**128**:127-161. DOI: 10.1016/bs.irn.2016.03.003

[25] Lesage F, Hibino H, Hudspeth AJ. Association of beta-catenin with the alpha-subunit of neuronal large-conductance Ca^{2+}-activated K^{+} channels. Proceedings of the National Academy of Sciences of the United States of America. 2004;**101**(2):671-675. DOI: 10.1073/pnas.0307681100

[26] Suzuki Y, Yamamura H, Ohya S, et al. Caveolin-1 facilitates the direct coupling between large conductance Ca^{2+}-activated K^{+} (BKCa) and Cav1.2 Ca^{2+} channels and their clustering to regulate membrane excitability in vascular myocytes. The Journal of Biological Chemistry. 2013;**288**(51):36750-36761. DOI: 10.1074/jbc.M113.511485

[27] Nelson MT, Cheng H, Rubart M, et al. Relaxation of arterial smooth muscle by calcium sparks. Science. 1995;**270**(5236):633-637

[28] Humphries ESA, Kamishima T, Quayle JM, et al. Calcium/calmodulin-dependent kinase 2 mediates Epac-induced spontaneous transient outward

currents in rat vascular smooth muscle. The Journal of Physiology. 2017;**595**(18):6147-6164. DOI: 10.1113/jp274754

[29] Lu T, Chai Q, Yu L, et al. Reactive oxygen species signaling facilitates FOXO-3a/FBXO-dependent vascular BK channel β1 subunit degradation in diabetic mice. Diabetes. 2012;**61**(7):1860-1868. DOI: 10.2337/db11-1658

[30] Zhang DM, He T, Katusic ZS, et al. Muscle-specific F-box only proteins facilitate BK channel β1 subunit downregulation in vascular smooth muscle cells of diabetes mellitus. Circulation Research. 2010;**107**(12):1454-1459. DOI: 10.1161/CIRCRESAHA.110.228361

[31] Lu T, Sun X, Li Y, et al. Role of Nrf2 signaling in the regulation of vascular BK channel beta1 subunit expression and BK channel function in high-fat diet-induced diabetic mice. Diabetes. 2017;**66**(10):2681-2690. DOI: 10.2337/db17-0181

[32] Fernandez-Fernandez JM et al. Gain-of-function mutation in the KCNMB1 potassium channel subunit is associated with low prevalence of diastolic hypertension. The Journal of Clinical Investigation. 2004;**113**(7):1032-1039. DOI: 10.1172/JCI20347

[33] Zhai X, Leo MD, Jaggar JH. Endothelin-1 stimulates vasoconstriction through Rab11A serine 177 phosphorylation. Circulation Research. 2017;**121**(6):650-661. DOI: 10.1161/circresaha.117.311102

[34] Leo MD, Zhai X, Yin W, et al. Impaired trafficking of beta1 subunits inhibits BK channels in cerebral arteries of hypertensive rats. Hypertension. 2018;**72**(3):765-775. DOI: 10.1161/HYPERTENSIONAHA.118.11147

[35] Rusko J, Tanzi F, van Breemen C, et al. Calcium-activated potassium channels in native endothelial cells from rabbit aorta: Conductance, Ca^{2+} sensitivity and block. The Journal of Physiology. 1992;**455**:601-621

[36] Naik JS, Osmond JM, Walker BR, et al. Hydrogen sulfide-induced vasodilation mediated by endothelial TRPV4 channels. American Journal of Physiology. Heart and Circulatory Physiology. 2016;**311**(6):H1437-H1444. DOI: 10.1152/ ajpheart.00465.2

[37] Sandow SL, Grayson TH. Limits of isolation and culture: Intact vascular endothelium and BKCa. American Journal of Physiology. Heart and Circulatory Physiology. 2009;**297**(1):H1-H7. DOI: 10.1152/ajpheart.00042.2

[38] Riddle MA, Hughes JM, Walker BR. Role of caveolin-1 in endothelial BKCa channel regulation of vasoreactivity. American Journal of Physiology—Cell Physiology. 2011;**301**(6):C1404-C1414. DOI: 10.1152/ajpcell.00013.2011

[39] Feher A, Rutkai I, Beleznai T, et al. Caveolin-1 limits the contribution of BK(Ca) channel to EDHF-mediated arteriolar dilation: Implications in diet-induced obesity. Cardiovascular Research. 2010;**87**(4):732-739. DOI: 10.1093/cvr088

[40] Bondarenko AI, Panasiuk O, Drachuk K, et al. The quest for endothelial atypical cannabinoid receptor: BKCa channels act as cellular sensors for cannabinoids in in vitro and in situ endothelial cells. Vascular Pharmacology. 2018;**102**:44-55. DOI: 10.1016/j.vph.2018.01.004

[41] Bondarenko AI, Panasiuk O, Okhai I, et al. Direct activation of Ca(2+) and voltage-gated potassium channels of large conductance by anandamide in endothelial cells does not support the presence of endothelial atypical cannabinoid receptor. European Journal of Pharmacology. 2017;**805**:14-24. DOI: 10.1016/j.ejphar.2017.03.038

[42] Cui M, Qin G, Yu K, et al. Targeting the small- and intermediate-conductance Ca^{2+}-activated potassium channels: The drug-binding pocket at the channel/calmodulin interface. Neuro-Signals. 2015;**22**(2):65-78. DOI: 10.1159/000367896

[43] Tang YR, Yang WW, Wang Y, et al. Estrogen regulates the expression of small-conductance Ca-activated K^+ channels in colonic smooth muscle cells. Digestion. 2015;**91**(3):187-196. DOI: 10.1159/000371544

[44] Tharp DL, Wamhoff BR, Turk JR, et al. Upregulation of intermediate-conductance Ca^{2+}-activated K^+ channel (IKCa1) mediates phenotypic modulation of coronary smooth muscle. American Journal of Physiology. Heart and Circulatory Physiology. 2006;**291**(5):H2493-H2503. DOI: 10.1152/ajpheart.01254.2

[45] Gu M, Zhu Y, Yin X, et al. Small-conductance Ca^{2+}-activated K^+ channels: Insights into their roles in cardiovascular disease. Experimental & Molecular Medicine. 2018;**50**(4):23. DOI: 10.1038/s12276-018-004

[46] Piazza M, Taiakina V, Dieckmann T, et al. Structural consequences of calmodulin EF hand mutations. Biochemistry. 2017;**56**(7):944-956. DOI: 10.1021/acs.biochem.6b01296

[47] Kohler M, Hirschberg B, Bond CT, et al. Small-conductance, calcium-activated potassium channels from mammalian brain. Science. 1996;**273**(5282):1709-1714

[48] Adelman JP. SK channels and calmodulin. Channels (Austin, Tex.). 2016;**10**(1):1-6. DOI: 0.1080/ 19336950. 2015.1029688

[49] Toldi G, Munoz L, Herrmann M, et al. The effects of Kv1.3 and IKCa1 channel inhibition on cytokine production and calcium influx of T lymphocytes in rheumatoid arthritis and ankylosing spondylitis. Immunologic Research. 2016;**64**(2): 627-631. DOI: 10.1007/s12026-015-868

[50] Chen YJ, Nguyen HM, Maezawa I, et al. The potassium channel KCa3.1 constitutes a pharmacological target for neuroinflammation associated with ischemia/reperfusion stroke. Journal of Cerebral Blood Flow and Metabolism. 2016;**36**(12):2146-2161. DOI: 10.1177/0271678X15611

[51] Feske S, Wulff H, Skolnik EY. Ion channels in innate and adaptive immunity. Annual Review of Immunology. 2015;**33**:291-353. DOI: 10.1146/annurev-immunol-032414-112212

[52] Yang H, Li X, Liu Y, et al. Crocin improves the endothelial function regulated by Kca3.1 through ERK and Akt signaling pathways. Cellular Physiology and Biochemistry. 2018;**46**(2):765-780. DOI: 10.1159/000488735

[53] Nguyen HM, Singh V, Pressly B, et al. Structural insights into the atomistic mechanisms of action of small molecule inhibitors targeting the KCa3.1 channel pore. Molecular Pharmacology. 2017;**91**(4):392-402. DOI: 10.1124/mol.116.108068

[54] Joiner WJ, Khanna R, Schlichter LC, et al. Calmodulin regulates assembly and trafficking of SK4/IK1 Ca^{2+}-activated K^+ channels. The Journal of Biological Chemistry. 2001;**276**(41):37980-37985. DOI: 10.1074/jbc.M104965200

[55] Garneau L, Klein H, Lavoie MF, et al. Aromatic-aromatic interactions between residues in KCa3.1 pore helix and S5 transmembrane segment control the channel gating process. The Journal of General Physiology. 2014;**143**(2):289-307. DOI: 10.1085/jgp.201311097

[56] Brasen JC, de Wit C, Sorensen CM. Myoendothelial coupling through Cx40 contributes to EDH-induced vasodilation in murine renal arteries: Evidence from experiments and modelling. Acta Physiologica (Oxford, England). 2018;**222**(1):e12906. DOI: 10.111/apha.12906

[57] Garland CJ. Compromised vascular endothelial cell SK(Ca) activity: A fundamental aspect of hypertension? British Journal of Pharmacology. 2010;**160**(4):833-835. DOI: 10.1111/j.1476-5381.2010.00692.x

[58] Desai KM, Gopalakrishnan V, Hiebert LM, et al. EDHF-mediated rapid restoration of hypotensive response to acetylcholine after chronic, but not acute, nitric oxide synthase inhibition in rats. European Journal of Pharmacology. 2006;**546**(1-3):120-126. DOI: 10.1016/j.ejphar.2006.06

[59] Yang G, Wu L, Jiang B, et al. H2S as a physiologic vasorelaxant: Hypertension in mice with deletion of cystathionine gamma-lyase. Science. 2008;**322**(5901):587-590. DOI: 10.1126/science.1162667

[60] Mustafa AK, Sikka G, Gazi SK, et al. Hydrogen sulfide as endothelium-derived hyperpolarizing factor sulfhydrates potassium channels. Circulation Research. 2011;**109**(11):1259-1268. DOI: 10.1161/CIRCRESAHA. 111.240242

[61] Wang M, Hu Y, Fan Y, et al. Involvement of hydrogen sulfide in endothelium-derived relaxing factor-mediated responses in rat cerebral arteries. Journal of Vascular Research. 2016;**53**(3-4):172-185. DOI: 10.1159/000448712

[62] Cheng Z, Shen X, Jiang X, et al. Hyperhomocysteinemia potentiates diabetes-impaired EDHF-induced vascular relaxation: Role of insufficient hydrogen sulfide. Redox Biology. 2018;**16**:215-225. DOI: 10.1016/j.redox.2018.02.0

[63] Behringer EJ, Segal SS. Tuning electrical conduction along endothelial tubes of resistance arteries through Ca($^{2+}$)-activated K(+) channels. Circulation Research. 2012;**110**(10):1311-1321. DOI: 10.1161/CIRCRESAHA. 111.262592

[64] Behringer EJ, Shaw RL, Westcott EB, et al. Aging impairs electric conduction along endothelium of resistance arteries through enhanced calcium-activated K^+ channel activation. Arteriosclerosis, Thrombosis, and Vascular Biology. 2013;**33**(8):1892-1901. DOI: 10.1161/ATVBAHA.113.301514

[65] Baukrowitz T, Schulte U, Oliver D, et al. PIP2 and PIP as determinants for ATP inhibition of KATP channels. Science. 1998;**282**(5391):1141-1144

[66] Martin GM, Yoshioka C, Rex EA, et al. Cryo-EM structure of the ATP-sensitive potassium channel illuminates mechanisms of assembly and gating. eLife. 2017;**6**:e24149. DOI: 10.7554/eLife.24149

[67] Liu SY, Tian HM, Liao DQ, et al. The effect of gliquidone on KATP channels in pancreatic beta-cells, cardiomyocytes, and vascular smooth muscle cells. Diabetes Research and Clinical Practice. 2015;**109**(2):334-339. DOI: 10.1016/j.diabres.2015.05.036

[68] Li CG, Cui WY, Wang H. Sensitivity of KATP channels to cellular metabolic disorders and the underlying structural basis. Acta Pharmacologica Sinica. 2016;**37**(1):134-142. DOI: 10.1038/aps.2015.134

[69] Nkanu EE, Owu DU, Osim EE. Altered potassium ion channel function as a possible mechanism of increased blood pressure in rats fed thermally oxidized palm oil diets. Journal of Dietary Supplements.

2018;**15**(4):431-444. DOI: 10.1080/19390211.2017.1350248

[70] Duling BR, Berne RM. Longitudinal gradients in periarteriolar oxygen tension. A possible mechanism for the participation of oxygen in local regulation of blood flow. Circulation Research. 1970;**27**(5):669-678

[71] Ngo AT, Jensen LJ, Riemann M, et al. Oxygen sensing and conducted vasomotor responses in mouse cremaster arterioles in situ. Pflügers Archiv: European Journal of Physiology. 2010;**460**(1):41-53. DOI: 10.1007/s00424-010-0837-x

[72] Montanari A, Pela G, Musiari L, et al. Nitric oxide-angiotensin II interactions and renal hemodynamic function in patients with uncomplicated type 1 diabetes. American Journal of Physiology. Renal Physiology. 2013;**305**(1):F42-F51. DOI: 10.1152/ajprenal.00109.2013

[73] Ikenaga H, Bast JP, Fallet RW, et al. Exaggerated impact of ATP-sensitive K(+) channels on afferent arteriolar diameter in diabetes mellitus. Journal of the American Society of Nephrology. 2000;**11**(7):1199-1207

[74] Salomonsson M, Brasen JC, Sorensen CM. Role of renal vascular potassium channels in physiology and pathophysiology. Acta Physiologica (Oxford, England). 2017;**221**(1):14-31. DOI: 10.1111/apha.12882

[75] Jimenez-Perez L, Cidad P, Alvarez-Miguel I, et al. Molecular determinants of Kv1.3 potassium channels-induced proliferation. The Journal of Biological Chemistry. 2016;**291**(7):3569-3580. DOI: 10.1074/jbc.M115.678995

[76] Long SB, Campbell EB, Mackinnon R. Voltage sensor of Kv1.2: Structural basis of electromechanical coupling. Science. 2005;**309**(5736):903-908. DOI: 10.1126/science.1116270

[77] Lamothe SM, Hogan-Cann AE, Li W, et al. The N terminus and transmembrane segment S1 of Kv1.5 can coassemble with the rest of the channel independent of the S1–S2 linkage. Journal of Biological Chemistry. 2018:A118-A4065. DOI: 10.1074/jbc.ra118.004065

[78] Albarwani S, Nemetz LT, Madden JA, et al. Voltage-gated K^+ channels in rat small cerebral arteries: Molecular identity of the functional channels. The Journal of Physiology. 2003;**551**(Pt 3):751-763. DOI: 10.1113/jphysiol.2003.040014

[79] Joseph BK, Thakali KM, Moore CL, et al. Ion channel remodeling in vascular smooth muscle during hypertension: Implications for novel therapeutic approaches. Pharmacological Research. 2013;**70**(1):126-138. DOI: 10.1061/j.phrs.2013.01.008

[80] Cox RH. Changes in the expression and function of arterial potassium channels during hypertension. Vascular Pharmacology. 2002;**38**(1):13-23

[81] Li Z, Lu N, Shi L. Exercise training reverses alterations in Kv and BKCa channel molecular expression in thoracic aorta smooth muscle cells from spontaneously hypertensive rats. Journal of Vascular Research. 2015;**51**(6): 447-457. DOI: 10.1159/000369928

[82] Goodwill AG, Noblet JN, Sassoon D, et al. Critical contribution of KV1 channels to the regulation of coronary blood flow. Basic Research in Cardiology. 2016;**111**(5):56. DOI: 10.1007/s00395-016-0575-0

[83] Ohanyan V, Yin L, Bardakjian R, et al. Requisite role of Kv1.5 channels in coronary metabolic dilation novelty and significance. Circulation Research. 2015;**117**(7):612-621. DOI: 10.1161/CIRCRESAHA.115.306642

[84] Yeung SY, Pucovsky V, Moffatt JD, et al. Molecular expression and

pharmacological identification of a role for K(v)7 channels in murine vascular reactivity. British Journal of Pharmacology. 2007;**151**(6):758-770. DOI: 10.1038/sj.bjp.0707284

[85] Khammy MM, Kim S, Bentzen BH, et al. 4-Aminopyridine: A pan voltage-gated potassium channel inhibitor that enhances Kv 7.4 currents and inhibits noradrenaline-mediated contraction of rat mesenteric small arteries. British Journal of Pharmacology. 2018;**175**(3):501-516. DOI: 101111/bph.14097

[86] Carr G, Barrese V, Stott JB, et al. MicroRNA-153 targeting of KCNQ4 contributes to vascular dysfunction in hypertension. Cardiovascular Research. 2016;**112**(2):581-589. DOI: 10.1093/cvr/cvw177

Chapter 3

L-Arginine/Nitric Oxide Pathway and KCa Channels in Endothelial Cells: A Mini-Review

Marcelo González and José Carlos Rivas

Abstract

The endothelium is an organ with a key role in the maintenance of cardiovascular health through the regulation of vascular tone, vascular resistance, blood flow, and arterial pressure. These functions are related with the synthesis and release of vasoactive molecules, mainly vasodilators like nitric oxide (NO) and endothelium-derived hyperpolarizing factor (EDHF). Both factors are released and diffused from endothelial cells to the smooth muscle cells, where there is a subsequent activation of signaling pathways that finally decrease the intracellular calcium to induce the vascular relaxation. The study of the molecular mechanisms that underlie the endothelial function still is in development, but from the evidence obtained from the endothelial cells *in vitro* studies are possible to partially describe the pathways to regulate the physiological endothelial function and the disturbances in pathological conditions. In this mini-review, we describe the main mechanisms for NO synthesis and the role of potassium channels related with EDHF. We include schemes and graphical summaries for better understanding of the molecular regulation of vascular tone in the human cardiovascular system.

Keywords: L-arginine, nitric oxide, potassium channels, endothelium

1. Characteristics of the endothelium

The endothelial cells (ECs) have mesenchymal origin, length of 25–50 μm and form a flat epithelium called endothelium. The endothelium in a human adult is composed of approximately 1–6 × 10^{13} cells, constituting an organ that weighs approximately 1 kg and covers a surface area of approximately 1–7 m^2 [1]. For decades, the endothelium was considered as a simple barrier between blood and the rest of the body's tissues. However, since the early 1980s, this vision changed radically [2] and, today, the endothelium is considered a true organ that fulfills multiple functions in the physiology and pathophysiology of vascular system, including autocrine, paracrine, and endocrine actions and the regulation of coagulation and fibrinolysis processes [3].

One of the most important functions of endothelial cells is their participation in the regulation of vascular tone. In the classic article of Furchgott and Zawadzki in 1980, it was demonstrated that the presence of the endothelium is essential for the vasodilator effect induced by acetylcholine in isolated blood vessels pre-constricted with norepinephrine. In those years, it was proposed that the vasodilation was

IntechOpen

produced through a factor that was released by the endothelium in response to agonists [4]. This factor was called the endothelial-derived relaxing factor (EDRF) [5]. Between 1986 and 1990, it was concluded that this factor corresponded to nitric oxide (NO) [6, 7]. The endothelium responds to mechanical stimuli such as pressure and flow stress ("shear stress"), hormonal stimuli, and vasoactive substances that regulate the vascular tone. The endothelial cells release molecules that regulate vasomotor function, inflammation, and hemostasis. Vasodilators agents include NO, prostacyclin, and endothelium-derived hyperpolarizing factor (EDHF). Vasoconstrictors agents include endothelin 1, angiotensin II, thromboxane A2, and reactive oxygen-derived species (ROS). Inflammatory mediators include NO, intercellular adhesion molecule 1 (ICAM-1), vascular adhesion molecule 1 (VCAM-1), E-selectin, and NFκB (**Figure 1** [8]).

Since the discovery of NO, the mechanisms of endothelial cell activation and endothelial dysfunction have been studied. In this way, the quiescent endothelial cells express a vasodilator, anticoagulant, and anti-adhesive phenotype, while the activated endothelial cell expresses procoagulant, pro-adhesive, and vasoconstrictive properties [9]. It has been considered that the decrease in the capacity of the vascular endothelium to stimulate vasodilation generates endothelial dysfunction, a phenomenon that is observed in several pathological conditions such as hypertension, hypercholesterolemia, diabetes mellitus, hyperhomocysteinemia, chronic kidney failure, chronic heart failure, etc. Although the molecular basis for endothelial dysfunction is not fully understood, numerous studies point to decreased biosynthesis and/or NO activity as a central mechanism [10–13].

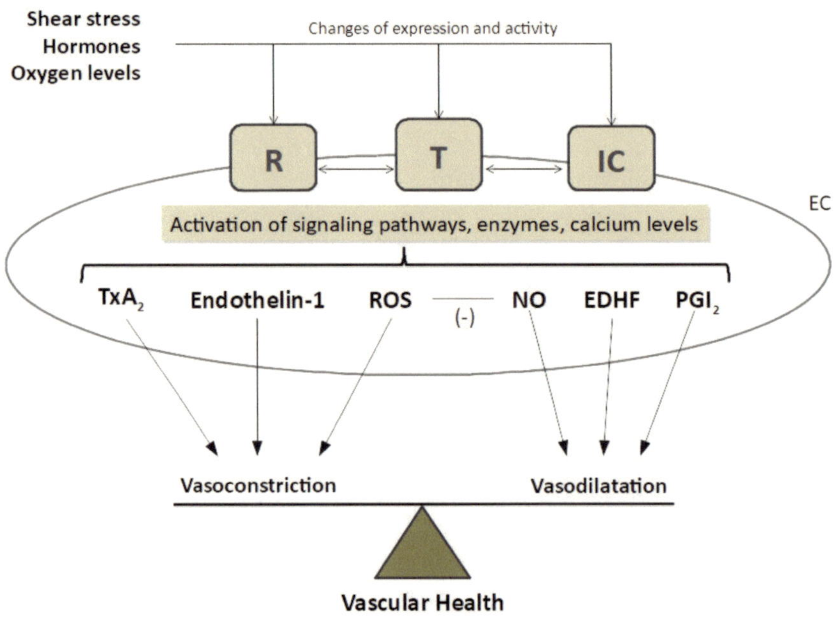

Figure 1.
Vascular tone regulation. The vascular tone is partially regulated by the local factors secreted by endothelial cells (ECs) in response to physical factors like shear stress and humoral and chemical factors like hormones and oxygen levels. The changes in blood flow are detected by membrane proteins, mainly receptors (Rs), transporters (Ts), and ion channels (ICs). There is a network connecting the activities of these proteins through signaling pathways that induce the release of different mediators like thromboxane A2 (TxA2), endothelin 1, reactive oxygen species (ROS), nitric oxide (NO), endothelium-derived hyperpolarizing factor (EDHF), or prostacyclin (PGI2), among others. The equilibrium between the vasoconstrictors and vasodilators factors maintains the endothelial function and vascular health.

2. Synthesis of nitric oxide in the endothelium

NO is synthesized from the semi-essential cationic amino acid L-arginine, which must be transported from the extracellular space into the endothelial cell by a family of cationic amino acid transporters (CATs) [14]. This amino acid is the substrate in a reaction where the metabolic product corresponds to L-citrulline in an equimolar proportion with the coproduct NO [15, 16]. This reaction is catalyzed by the enzyme NO synthase (NOS), which can be classified into their constitutive forms (cNOS) and their inducible form (iNOS) [17]. The cNOS includes the endothelial isoform (eNOS) and the neuronal isoform (nNOS), both producing NO in short bursts at low concentrations (nM) and in a calcium-dependent manner to fulfill the physiological functions of NO. The physiological activity of eNOS is dependent on several cofactors and is regulated by signaling pathways that induce phosphorylation in different sites for activation (serine 1177) or inhibition (threonine 495) [17]. NO diffuses from endothelial cells to smooth muscle cells (SMCs) and activates the soluble guanylate cyclase (sGC) pathway, to reduce the intracellular calcium and induce vasodilation (**Figure 2**). iNOS is mainly expressed in cells that participate in the inflammatory response after induction by cytokines and other inflammatory mediators, producing NO in high concentrations (μM) and independently of calcium [18–20].

The availability of NO in vivo is regulated by a combination of NO synthesis and inactivation. The decrease in the availability of NO may be due to a lower expression

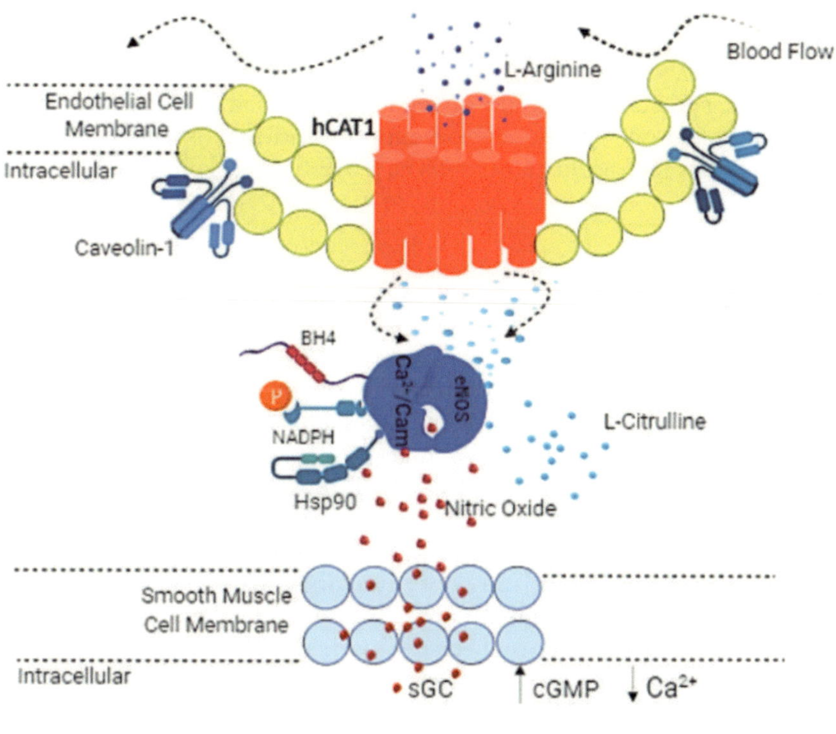

Figure 2.
L-arginine transport and nitric oxide synthesis in endothelial cells. hCAT-1 is a protein expressed in plasma membrane of endothelial cells, mainly in plasma membrane invagination called caveolae. The L-arginine enters to the cell from blood and is used by eNOS to synthesize L-citrulline and nitric oxide (NO). The eNOS needs different cofactors to maintain its function, which include tetrahydrobiopterin (BH_4), nicotidamine adenine dinuclotide phosphate (NADPH), and heat shock protein 90 (Hsp90). Nitric oxide diffuses through the cell membranes and enters the smooth muscle cells to activate the soluble guanlylate cyclase (sGC). The sGC synthesizes cyclic GMP (cGMP), which activates protein kinase G and, after subsequent steps, the intracellular calcium decreases to induce the vasodilation.

or activity of eNOS, as a result of the action of endogenous and exogenous inhibitors or due to the lower availability of the substrate L-arginine [8, 14]. The availability of NO can also be diminished by the rapid reaction between NO and reactive oxygen-derived species (ROS) [13].

3. Reactive oxygen-derived species (ROS) in endothelium

Endothelial cells generate ROS, including the superoxide radical ($O_2^{\cdot -}$), hydrogen peroxide (H_2O_2), peroxynitrite ($ONOO^-$), hydroxyl radical ($^{\cdot}OH$), among others [15, 16]. In endothelial cells, the main sources of ROS are the enzymatic complex xanthine oxidoreductase (XOR) [17], the complex of membrane nicotinamide adenine dinucleotide phosphate oxidase (NADPH oxidase) [18], eNOS itself when it is "uncoupled" due to lack of tetrahydrobiopterin (BH_4) or L-arginine [19], mitochondrial cytochromes [20], and hemoglobin [21].

Among all endothelial ROS sources, NADPH oxidases are enzymes whose primary function is the generation of ROS and they play an important role in redox signaling [22]. On the other hand, the activity of NADPH oxidase can cause the uncoupling of eNOS by the oxidative degradation of BH_4, leading to the eNOS-dependent synthesis of $O_2^{\cdot -}$ and detriment of the synthesis of NO [18, 23]. Once $O_2^{\cdot -}$ is synthesized, it can act as a precursor to other ROS due to its use by superoxide dismutase (SOD) to generate H_2O_2 that has greater stability and capacity to cross biological membranes, and it therefore can act as a modulator of signal transduction pathways [24]. Furthermore, $O_2^{\cdot -}$ reacts quickly with NO to generate $ONOO^-$, a powerful oxidizing agent that causes DNA fragmentation and lipid oxidation [25].

It is currently postulated that the mechanism by which $O_2^{\cdot -}$ "kidnaps" NO would play a central role in the development of endothelial dysfunction that is seen in pathologies such as diabetes mellitus [26–28], preeclampsia [29, 30], and hypertension [31].

4. L-arginine transport in human fetal endothelium

The amino acid L-arginine is taken up by endothelial cells through the transporter systems y^+, y^+L, $b^{0,+}$, and $y\,B^{0,+}$ [32–35]. Of these systems, there are two that have been described in HUVEC, that is, y^+ system [36–38] and y^+L system [39]. The y^+ system family is currently known to include at least five cationic amino acid transporters (CATs) called CAT-1, CAT-2A, CAT-2B, CAT-3, and CAT-4. CAT-1 is expressed ubiquitously, CAT-2A and CAT-3 are constitutively expressed in liver and brain, respectively, while CAT-2B is induced in a variety of cell types in response to bacterial endotoxins and pro-inflammatory cytokines [40, 41]. CAT-4 corresponds to a cDNA sequence with 41–42% identity with the other members of the CATs family, but its transport activity has not yet been determined [32, 34, 35]. CAT-1, CAT-2B, and CAT-3 are characterized by high affinity to the substrate (K_m = 100–400 μM) and independency of Na^+, while CAT-2A has low affinity for cationic amino acids (K_m = 2–5 mM). Two members of CATs have been reported to be expressed in HUVEC, that is, hCAT-1 and hCAT-2B, while hCAT-2A and hCAT-3 transporters have not been detected in this cell type [34, 36–39] (**Table 1**). Although the hCAT-1 and hCAT-2B transporters have similar kinetic characteristics, it is possible to differentiate them by their different sensitivities to L-lysine trans-stimulation. In *Xenopus laevis* oocytes injected with hCAT-1 and hCAT-2B mRNA, L-lysine increases L-arginine transport by 9.8-fold and 1.8-fold, respectively [42]. Thus, for L-lysine trans-stimulation assays in HUVEC, it has been possible to determine that

Gene	Protein	K_m (μM)	Distribution
SLC7A1	CAT-1	70–250	All tissues except liver and lacrimal gland
SLC7A2	CAT-2A	2.2–5.2	Liver, skeletal muscle, and pancreas
SLC7A2	CAT-2B	38–380	Endothelium, and inducible in several tissues
SLC7A3	CAT-3	40–120	Thymus, ovary, testes, and brain
SLC7A4	CAT-4	—	Brain, testes, and placenta

Proteins CATs are coded in different genes (except CAT2A and 2B, same gene), have different kinetic constants for the transport of L-arginine (K_m) and distribution in tissues.

Table 1.
CATs' family members.

the hCAT-1 transporter accounts for 60–80% of the total uptake of L-arginine in physiological conditions [36–38]. The importance of the hCAT-1 transporter in NO synthesis has been confirmed through a transgenic mouse model that overexpresses the protein exclusively in the endothelium. Aortic rings obtained from these transgenic mice have a higher sensitivity to relaxation in response to acetylcholine compared to native mice, while endothelial cell cultures obtained from these animals, that overexpress hCAT-1, exhibit a greater NO synthesis [43].

5. Regulation of the expression of hCAT-1

Regarding the gene organization of CAT transporters, it is known that the *SLC7* family is phylogenetically composed of two subfamilies formed by cationic amino acid transporters (CATs) and glycoprotein-associated amino acid transporters (HATs). The cationic amino acid transporter family is encoded by the *SLC7A* (1–4) genes and corresponds to proteins with 14 transmembrane domains [44]. Specifically, the gene that encodes the hCAT-1 protein corresponds to *SLC7A1* whose open reading frame is formed by 11 exons and 10 introns. The gene is located on chromosome 13q12-13q14 [45].

Among the genes encoding CAT-1 in rat, mouse and human have common characteristics: the promoter region lacks TATA box, and they have multiple binding sites for the transcription factor specific protein 1 (Sp1) and they have an extensive 3′ non-translatable region (3′UTR) that could perform functions in the regulation of mRNA stability or in translation [46–49]. In rats, stress by amino acids deprivation induces an increase in the rCAT-1 mRNA expression by a mechanism related to increased mRNA stability [46]. This increased mRNA stability would be related to the presence of a regulatory region within the 3′UTR sequence of the gene [47]. Subsequent experiments have shown that the effect of amino acids deprivation on rCAT-1 expression would depend on both transcriptional [48] and posttranscriptional mechanisms [50].

In humans, it is known that insulin increases leg blood flow in healthy subjects via stimulation of endothelial NO synthase (eNOS) [51]. Insulin also increases the synthesis and release of NO and release in primary cultures of HUVEC [38, 52]. Biological effects of insulin involve activation of several transcription factors, including Sp1 in several cell types [53, 54]. Insulin increases Sp1 nuclear protein abundance and its binding to a proximal region (−177 and −105 bp from ATG) of the *SLC7A1* promoter containing four consensus sequences for Sp1 [55]. Interestingly, in patients with essential hypertension, a reduction of *SLC7A1* transcriptional activity due to reduced Sp1 activity in the promoter region has been reported [12]. So, the transcriptional regulation of *SLC7A1* is relevant for cardiovascular physiology, and the reduction of the promoter activity of this gene could be associated with cardiovascular disease (CVD).

On the other hand, the first intron of *SLC7A1* may play a bifunctional role in regulating the *SLC7A1* transcriptional activity by the binding of the purine-rich element binding protein A (Pur alpha) in physiological conditions and by activating the transcription factor 4 (ATF4) in endoplasmic reticulum stress or by decreasing the *SLC7A1* transcriptional activity by the C/EBP homologous protein 10 (CHOP) binding in C6 rat glioma cells [56].

For the physiological regulation of hCAT-1 activity, both transcriptional regulation of *SLC7A1* and/or posttranscriptional regulation of *SLC7A1* transcript are relevant for the protein expression and L-arginine transport [55]. Insulin increases the expression of *SLC7A1* gene due to an increased transcriptional activity, most likely due to higher Sp1 activity. So, hCAT-1 expression and activity are regulated by insulin in endothelium, suggesting that in insulin resistance there is a reduction of L-arginine transport and NO synthesis that contributes to endothelial dysfunction and cardiovascular diseases.

6. High D-glucose and expression and activity of L-arginine/NO pathway

Hyperglycemia and diabetes mellitus are pathological conditions associated with fetal endothelial dysfunction [55] and type 2 diabetes mellitus (T2DM) [57] or cardiovascular disease (CVD) [58]. CVD in patients with diabetes mellitus is associated with the generation of ROS.

High concentration of D-glucose (25 mM) increases L-arginine transport and cGMP accumulation in endothelium in a similar manner to that observed in HUVEC from pregnancies with gestational diabetes [33, 59]. Increased L-arginine transport in response to incubation with high D-glucose has been related to increased mRNA levels for the hCAT-1 and eNOS activity in HUVEC [60]. In human aortic endothelial cells, prolonged incubation (7 days) with 25 mM D-glucose induces a decrease in eNOS activity (determined by nitrite content), protein abundance, and mRNA level. This effect is associated with a decrease in eNOS promoter activity [61]. In bovine aortic endothelial cells (BAECs), there is a lower production of insulin-induced NO when the cells were incubated with high extracellular concentration of D-glucose, an effect that seems to depend on a signaling pathway that involves to the type 1 insulin receptor (IR-1), phosphatidyl inositol 3 kinase, and the inhibitor of nuclear factor kappa-B kinase [62]. On the other hand, the increase of cGMP production induced by high D-glucose in HUVEC is blocked by incubating the cells with 1 nM insulin [63]. Incubation with 1 nM insulin (8 h) has been shown in this same cell type to be sufficient to block the effect that D-glucose has on the decreased transport of adenosine [64], an important vasoactive nucleoside [65].

In HUVEC, high extracellular D-glucose increases L-arginine transport, NO synthesis, and $O_2^{\cdot -}$ generation through eNOS and NADPH oxidase activation. Additionally, high D-glucose increased the contractile response in the human umbilical vein. Insulin reversed these effects of high D-glucose, leading to normal hCAT-1 expression, NO synthesis, ROS generation, and vascular tone. Insulin acts like antioxidant molecules (like tempol, ascorbic acid) to restore high D-glucose-increased oxidative stress in the fetoplacental vascular bed [66]. High D-glucose increases L-arginine transport, likely resulting from higher hCAT-1 expression and protein abundance in the plasma membrane. This mechanism could be an adaptive response of HUVEC to higher ROS generation from high D-glucose-activated

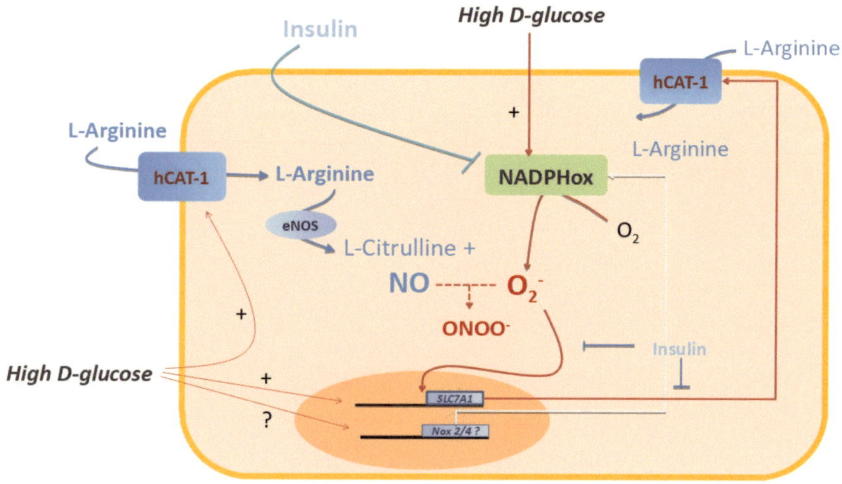

Figure 3.
Endothelial dysfunction induced by high D-glucose and protection by insulin in HUVEC. Exposure of HUVEC to high D-glucose leads to an increase (↑) in the plasma membrane abundance of the human cationic amino acid transporter 1 (hCAT-1) and higher L-arginine uptake. High D-glucose activates NADPH oxidase, leading to higher generation of ROS, including $O_2^{\cdot -}$. Insulin restores ROS and $O_2^{\cdot -}$ generation to values in cells exposed to 5 mM D-glucose (normal), resulting in the restoration of hCAT-1-mediated L-arginine transport and nitric oxide (NO) synthesis. High D-glucose and insulin also activate the SLC7A1 *promoter region (coding for hCAT-1) up to −650 bp from the ATG via a mechanism involving ROS and $O_2^{\cdot -}$ generation. In addition, insulin restores hCAT-1 protein abundance and its distribution in the cells via an NADPH oxidase-independent mechanism (data from González et al. [66]).*

NADPH oxidase. In parallel, high D-glucose increased NO synthesis. Insulin reversed the high D-glucose-mediated alterations in L-arginine transport involving the modulation of *SLC7A1* gene expression, leading to altered umbilical vein reactivity. Modulation of hCAT-1 expression and activity by insulin is the key to maintaining umbilical vein tone and endothelial function in physiologic and pathophysiological conditions (**Figure 3**) [66].

7. Role of potassium channels in endothelial function

Another important mechanism that regulates the endothelial function is the activity of ion channels that modulate the cell membrane potential. The calcium-activated potassium channels (KCa) have been shown to be relevant to induce the necessary hyperpolarization to stimulate the relaxation of vascular smooth muscle cells (related with EDHF). In systemic circulation, large conductance KCa (BKCa) channels have been shown preferentially expressed in VSMC, meanwhile small (SKCa) and intermediate (IKCa) conductance KCa are preferentially expressed in endothelium [67, 68]. However, potassium currents inhibited by iberiotoxin (BKCa inhibitor) have been described in HUVEC stimulated by sildenafil or insulin [69]. In fact, insulin (10 nM) can directly activate native and recombinant BKCa currents in cell-attached patch-clamping experiments with a rapid effect that is MAPK-dependent when the hormone was added in the pipette [70]. There is evidence that insulin may induce endothelial cell hyperpolarization by modulating K channels activity [38, 71]. The insulin-induced relaxation in human placental veins (~368 μm diameter), pre-constricted with U46619, is a mechanism dependent on the BKCa channel activity. The co-incubation of vessels with genistein (tyrosine kinases inhibitor) and wortmannin (PI3K inhibitor) did not block the insulin's relaxation, and by contrast potentiated the insulin-induced

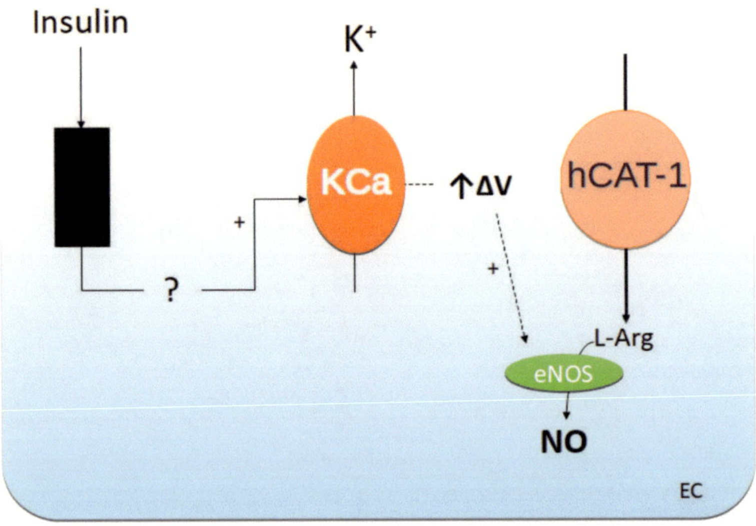

Figure 4.
Proposal of mechanism for KCa activation by insulin. Evidence obtained in endothelial cells (ECs) shows that insulin activates KCa (mainly BKCa) in a mechanism still not fully understood. The activation of KCa by insulin induces hyperpolarization (↑ΔV), leading to activation of eNOS for NO synthesis from L-arginine uptake by hCAT-1 (modified from Rojas et al. [71]).

vasodilation. Also, insulin decreased perfusion pressure (34 ± 3%) in the isolated cotyledon of normal placenta with a basal perfusion pressure of 64 ± 5 mmHg (or pre-constricted with U46619) [72]. The effects of insulin on BKCa activity are associated with evidences that show that the constriction induced by U46619 and H_2O_2 in placental vasculature is partially decreased with 10 nM insulin preincubation (10 min) in a mechanism totally dependent of BKCa activity [72]. Recently, it has been determined that insulin-mediated NO synthesis requires the participation of both IKCa/ BKCa channels and eNOS activity in HUVECs [71]. In the same cell type, insulin increased the open probability (NPo) of BKCa, associated with hyperpolarization in single cell analysis [69]. In human placental arteries, the relaxation induced by the NO donor, SNAP, is partially blocked by charybdotoxin (BKCa inhibitor) and almost totally blocked by charybdotoxin and ODQ (sGC inhibitor) [73]. Therefore, an extracellular stimulus that increases the NO availability activates a mechanism that involves sGC and BKCa activities [71]. These findings constitute evidence for postulating a new mechanism induced by insulin in human vasculature related with the physiological regulation of KCa activity for NO synthesis (**Figure 4**).

8. Final remarks

The relevance of the endothelium for cardiovascular physiology is well established, mainly by findings related to the capacity of endothelial cells to synthesize NO and regulate the plasma membrane potential of smooth muscle cells. **Figure 5** shows a graphical summary of the L-arginine/NO pathway in the human blood vessels that highlight the capacity of endothelial cells to respond to extracellular stimuli and translate the mechanical forces and endocrine signals to intracellular mechanisms leading to NO synthesis and activation of potassium channels. It is important to note that the subcellular distribution of hCAT-1 and eNOS is also relevant for endothelial cells function. In physiological state, hCAT-1 colocalizes with caveolin-1 in the plasma membrane caveolae in proximity to eNOS.

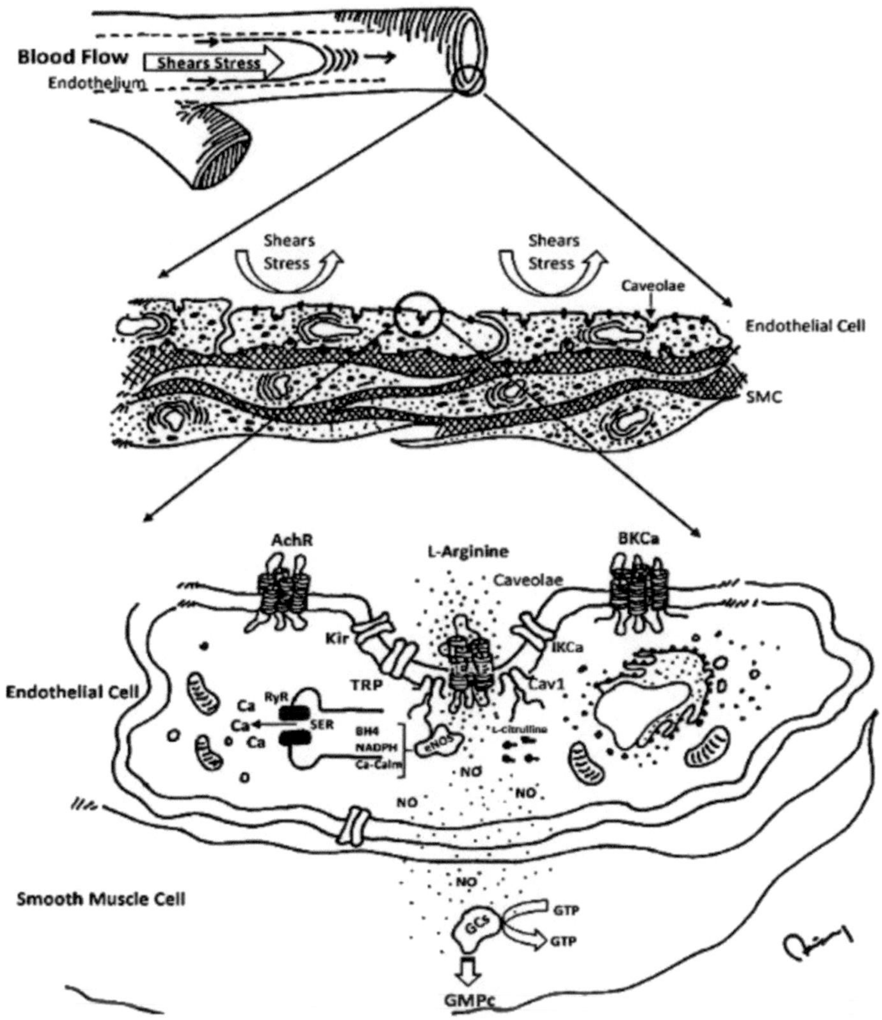

Figure 5.
Role of endothelium in the regulation of vascular tone. Endothelial cells, as a part of blood vessels walls, respond to mechanical stress induced by flow (shear stress) by activation of L-arginine/NO pathway to induce the NO release and relaxation of smooth muscle cells (SMCs). Subcellular localization of hCAT-1 in caveolae is relevant for its function, and the role of potassium channels (BKCa, mainly) has been recently described as important for endothelial cells function. The activity of the endothelium is regulated by different agonists like acetylcholine (Ach) through plasma membrane receptor (AchR) and others like insulin or serotonin, etc.

Acknowledgements

The authors would like to acknowledge the staff at Laboratorio de Investigación Materno-Fetal (LIMaF) and Department of Obstetrics and Gynecology from the Universidad de Concepción. The authors express special thanks to Alexandra Elbakyan for support the open science.

Author details

Marcelo González* and José Carlos Rivas
Laboratorio de Investigación Materno Fetal (LIMaF), Departamento de Obstetricia y Ginecología, Facultad de Medicina, Universidad de Concepción, Concepción, Chile

*Address all correspondence to: mgonzalezo@udec.cl

References

[1] Augustin HG, Kozian DH, Johnson RC. Differentiation of endothelial cells: Analysis of the constitutive and activated endothelial cell phenotypes. BioEssays: News and Reviews in Molecular, Cellular and Developmental Biology. 1994;**16**(12):901-906

[2] Fishman AP. Endothelium: A distributed organ of diverse capabilities. Annals of the New York Academy of Sciences. 1982;**401**:1-8

[3] Galley HF, Webster NR. Physiology of the endothelium. British Journal of Anaesthesia. 2004;**93**(1):105-113

[4] Furchgott RF, Zawadzki JV. The obligatory role of endothelial cells in the relaxation of arterial smooth muscle by acetylcholine. Nature. 1980;**288**(5789):373-376

[5] Griffith TM, Edwards DH, Lewis MJ, Newby AC, Henderson AH. The nature of endothelium-derived vascular relaxant factor. Nature. 1984;**308**(5960):645-647

[6] Ignarro LJ, Buga GM, Wood KS, Byrns RE, Chaudhuri G. Endothelium-derived relaxing factor produced and released from artery and vein is nitric oxide. Proceedings of the National Academy of Sciences of the United States of America. 1987;**84**(24):9265-9269

[7] Moncada S, Palmer RM, Higgs EA. The discovery of nitric oxide as the endogenous nitrovasodilator. Hypertension (Dallas, Tex.: 1979). 1988;**12**(4):365-372

[8] Endemann DH, Schiffrin EL. Endothelial dysfunction. Journal of the American Society of Nephrology: JASN. 2004;**15**(8):1983-1992

[9] Aird WC. Endothelium in health and disease. Pharmacological Reports. 2008;**60**(1):139-143

[10] De Meyer GR, Herman AG. Vascular endothelial dysfunction. Progress in Cardiovascular Diseases. 1997;**39**(4):325-342

[11] Kurowska EM. Nitric oxide therapies in vascular diseases. Current Pharmaceutical Design. 2002;**8**(3):155-166

[12] Yang Z, Kaye DM. Endothelial dysfunction and impaired L-arginine transport in hypertension and genetically predisposed normotensive subjects. Trends in Cardiovascular Medicine. 2006;**16**(4):118-124

[13] González M. Regulation of expression and activity of l-arginine transporters by nutrients and hormones: A focus in transcriptional mechanisms regulated by glucose and insulin. En: Patel VB, Preedy VR, Rajendram R, editores. L-Arginine in Clinical Nutrition (Nutrition and Health). Cham: Springer International Publishing; 2017. pp. 71-83. DOI: 10.1007/978-3-319-26009-9_6 [Accessed: 12 July 2020]

[14] Kalinowski L, Malinski T. Endothelial NADH/NADPH-dependent enzymatic sources of superoxide production: Relationship to endothelial dysfunction. Acta Biochimica Polonica. 2004;**51**(2):459-469

[15] Förstermann U. Oxidative stress in vascular disease: Causes, defense mechanisms and potential therapies. Nature Clinical Practice. Cardiovascular Medicine. 2008;**5**(6):338-349

[16] Valko M, Leibfritz D, Moncol J, Cronin MTD, Mazur M, Telser J. Free radicals and antioxidants in normal physiological functions and human disease. The International Journal of Biochemistry & Cell Biology. 2007;**39**(1):44-84

[17] Berry CE, Hare JM. Xanthine oxidoreductase and cardiovascular disease: Molecular mechanisms and pathophysiological implications. The Journal of Physiology. 2004;**555**(Pt 3): 589-606

[18] Dworakowski R, Alom-Ruiz SP, Shah AM. NADPH oxidase-derived reactive oxygen species in the regulation of endothelial phenotype. Pharmacological Reports. 2008;**60**(1):21-28

[19] Münzel T, Daiber A, Ullrich V, Mülsch A. Vascular consequences of endothelial nitric oxide synthase uncoupling for the activity and expression of the soluble guanylyl cyclase and the cGMP-dependent protein kinase. Arteriosclerosis, Thrombosis, and Vascular Biology. 2005;**25**(8):1551-1557

[20] Cadenas E, Sies H. The lag phase. Free Radical Research. 1998;**28**(6):601-609

[21] Balla J, Vercellotti GM, Nath K, Yachie A, Nagy E, Eaton JW, et al. Haem, haem oxygenase and ferritin in vascular endothelial cell injury. Nephrology, Dialysis, Transplantation: Official Publication of the European Dialysis and Transplant Association - European Renal Association. 2003;**18**(Suppl 5):v8-v12

[22] Lambeth JD. NOX enzymes and the biology of reactive oxygen. Nature Reviews. Immunology. 2004;**4**(3):181-189

[23] Antoniades C, Shirodaria C, Warrick N, Cai S, de Bono J, Lee J, et al. 5-methyltetrahydrofolate rapidly improves endothelial function and decreases superoxide production in human vessels: Effects on vascular tetrahydrobiopterin availability and endothelial nitric oxide synthase coupling. Circulation. 2006;**114**(11):1193-1201

[24] Li J-M, Shah AM. Endothelial cell superoxide generation: Regulation and relevance for cardiovascular pathophysiology. American Journal of Physiology. Regulatory, Integrative and Comparative Physiology. 2004;**287**(5):R1014-R1030

[25] Carr AC, McCall MR, Frei B. Oxidation of LDL by myeloperoxidase and reactive nitrogen species: Reaction pathways and antioxidant protection. Arteriosclerosis, Thrombosis, and Vascular Biology. 2000;**20**(7):1716-1723

[26] Rolo AP, Palmeira CM. Diabetes and mitochondrial function: Role of hyperglycemia and oxidative stress. Toxicology and Applied Pharmacology. 2006;**212**(2):167-178

[27] Hadi HAR, Suwaidi JA. Endothelial dysfunction in diabetes mellitus. Vascular Health and Risk Management. 2007;**3**(6):853-876

[28] Rask-Madsen C, King GL. More sugar, less blood vessels: Another piece in the puzzle of increased cardiovascular risk in diabetes. Arteriosclerosis, Thrombosis, and Vascular Biology. 2008;**28**(4):608-610

[29] Gu Y, Lewis DF, Zhang Y, Groome LJ, Wang Y. Increased superoxide generation and decreased stress protein Hsp90 expression in human umbilical cord vein endothelial cells (HUVECs) from pregnancies complicated by preeclampsia. Hypertension in Pregnancy. 2006;**25**(3):169-182

[30] Escudero C, Sobrevia L. A hypothesis for preeclampsia: Adenosine and inducible nitric oxide synthase in human placental microvascular endothelium. Placenta. 2008;**29**(6):469-483

[31] Harrison DG, Cai H, Landmesser U, Griendling KK. Interactions of angiotensin II with NAD(P)H oxidase,

oxidant stress and cardiovascular disease. Journal of the Renin-Angiotensin-Aldosterone System. 2003;**4**(2):51-61

[32] Mann GE, Yudilevich DL, Sobrevia L. Regulation of amino acid and glucose transporters in endothelial and smooth muscle cells. Physiological Reviews. 2003;**83**(1):183-252

[33] Sobrevia L, Mann GE. Dysfunction of the endothelial nitric oxide signalling pathway in diabetes and hyperglycaemia. Experimental Physiology. 1997;**82**(3):423-452

[34] Casanello P, Escudero C, Sobrevia L. Equilibrative nucleoside (ENTs) and cationic amino acid (CATs) transporters: Implications in foetal endothelial dysfunction in human pregnancy diseases. Current Vascular Pharmacology. 2007;**5**(1):69-84

[35] Sobrevia L, González M. A role for insulin on L-arginine transport in fetal endothelial dysfunction in hyperglycaemia. Current Vascular Pharmacology. 2009;**7**(4):467-474

[36] Casanello P, Sobrevia L. Intrauterine growth retardation is associated with reduced activity and expression of the cationic amino acid transport systems y^+/hCAT-1 and y^+/hCAT-2B and lower activity of nitric oxide synthase in human umbilical vein endothelial cells. Circulation Research. 2002;**91**(2):127-134

[37] Flores C, Rojas S, Aguayo C, Parodi J, Mann G, Pearson JD, et al. Rapid stimulation of L-arginine transport by D-glucose involves p42/44(mapk) and nitric oxide in human umbilical vein endothelium. Circulation Research. 2003;**92**(1):64-72

[38] González M, Flores C, Pearson JD, Casanello P, Sobrevia L. Cell signalling-mediating insulin increase of mRNA expression for cationic amino acid transporters-1 and -2 and membrane hyperpolarization in human umbilical vein endothelial cells. Pflügers Archiv. 2004;**448**(4):383-394

[39] Arancibia-Garavilla Y, Toledo F, Casanello P, Sobrevia L. Nitric oxide synthesis requires activity of the cationic and neutral amino acid transport system y^+L in human umbilical vein endothelium. Experimental Physiology. 2003;**88**(6):699-710

[40] Devés R, Boyd CA. Transporters for cationic amino acids in animal cells: Discovery, structure, and function. Physiological Reviews. 1998;**78**(2):487-545

[41] Palacín M, Estévez R, Bertran J, Zorzano A. Molecular biology of mammalian plasma membrane amino acid transporters. Physiological Reviews. 1998;**78**(4):969-1054

[42] Closs EI, Gräf P, Habermeier A, Cunningham JM, Förstermann U. Human cationic amino acid transporters hCAT-1, hCAT-2A, and hCAT-2B: Three related carriers with distinct transport properties. Biochemistry. 1997;**36**(21):6462-6468

[43] Yang Z, Venardos K, Jones E, Morris BJ, Chin-Dusting J, Kaye DM. Identification of a novel polymorphism in the 3'UTR of the L-arginine transporter gene SLC7A1: Contribution to hypertension and endothelial dysfunction. Circulation. 2007;**115**(10):1269-1274

[44] Verrey F, Closs EI, Wagner CA, Palacin M, Endou H, Kanai Y. CATs and HATs: The SLC7 family of amino acid transporters. Pflügers Archiv. 2004;**447**(5):532-542

[45] Hammermann R, Brunn G, Racké K. Analysis of the genomic organization of the human cationic amino acid transporters CAT-1, CAT-2 and CAT-4. Amino Acids. 2001;**21**(2):211-219

[46] Aulak KS, Liu J, Wu J, Hyatt SL, Puppi M, Henning SJ, et al. Molecular sites of regulation of expression of the rat cationic amino acid transporter gene. The Journal of Biological Chemistry. 1996;**271**(47):29799-29806

[47] Aulak KS, Mishra R, Zhou L, Hyatt SL, de Jonge W, Lamers W, et al. Post-transcriptional regulation of the arginine transporter Cat-1 by amino acid availability. The Journal of Biological Chemistry. 1999;**274**(43):30424-30432

[48] Fernandez J, Lopez AB, Wang C, Mishra R, Zhou L, Yaman I, et al. Transcriptional control of the arginine/lysine transporter, cat-1, by physiological stress. The Journal of Biological Chemistry. 2003;**278**(50):50000-50009

[49] Hatzoglou M, Fernandez J, Yaman I, Closs E. Regulation of cationic amino acid transport: The story of the CAT-1 transporter. Annual Review of Nutrition. 2004;**24**:377-399

[50] Lopez AB, Wang C, Huang CC, Yaman I, Li Y, Chakravarty K, et al. A feedback transcriptional mechanism controls the level of the arginine/lysine transporter cat-1 during amino acid starvation. The Biochemical Journal. 2007;**402**(1):163-173

[51] Steinberg HO, Baron AD. Vascular function, insulin resistance and fatty acids. Diabetologia. 2002;**45**(5):623-634

[52] Scherrer U, Randin D, Vollenweider P, Vollenweider L, Nicod P. Nitric oxide release accounts for insulin's vascular effects in humans. The Journal of Clinical Investigation. 1994;**94**(6):2511-2515

[53] Samson SL-A, Wong NCW. Role of Sp1 in insulin regulation of gene expression. Journal of Molecular Endocrinology. 2002;**29**(3):265-279

[54] Solomon SS, Majumdar G, Martinez-Hernandez A, Raghow R. A critical role of Sp1 transcription factor in regulating gene expression in response to insulin and other hormones. Life Sciences. 2008;**83**(9-10):305-312

[55] González M, Gallardo V, Rodríguez N, Salomón C, Westermeier F, Guzmán-Gutiérrez E, et al. Insulin-stimulated L-arginine transport requires SLC7A1 gene expression and is associated with human umbilical vein relaxation. Journal of Cellular Physiology. 2011;**226**(11):2916-2924

[56] Huang CC, Chiribau C-B, Majumder M, Chiang C-M, Wek RC, Kelm RJ, et al. A bifunctional intronic element regulates the expression of the arginine/lysine transporter Cat-1 via mechanisms involving the purine-rich element binding protein A (Pur alpha). The Journal of Biological Chemistry. 2009;**284**(47):32312-32320

[57] Damm P. Future risk of diabetes in mother and child after gestational diabetes mellitus. International Journal of Gynaecology and Obstetrics: The Official Organ of the International Federation of Gynaecology and Obstetrics. 2009;**104**(Suppl 1):S25-S26

[58] Brewster S, Zinman B, Retnakaran R, Floras JS. Cardiometabolic consequences of gestational dysglycemia. Journal of the American College of Cardiology. 2013;**62**(8):677-684

[59] Sobrevia L, Nadal A, Yudilevich DL, Mann GE. Activation of L-arginine transport (system y^+) and nitric oxide synthase by elevated glucose and insulin in human endothelial cells. The Journal of Physiology. 1996;**490**(Pt 3): 775-781

[60] Vásquez R, Farías M, Vega JL, Martin RS, Vecchiola A, Casanello P, et al. D-glucose stimulation of

L-arginine transport and nitric oxide synthesis results from activation of mitogen-activated protein kinases p42/44 and Smad2 requiring functional type II TGF-beta receptors in human umbilical vein endothelium. Journal of Cellular Physiology. 2007;**212**(3):626-632

[61] Srinivasan S, Hatley ME, Bolick DT, Palmer LA, Edelstein D, Brownlee M, et al. Hyperglycaemia-induced superoxide production decreases eNOS expression via AP-1 activation in aortic endothelial cells. Diabetologia. 2004;**47**(10):1727-1734

[62] Kim F, Tysseling KA, Rice J, Gallis B, Haji L, Giachelli CM, et al. Activation of IKKbeta by glucose is necessary and sufficient to impair insulin signaling and nitric oxide production in endothelial cells. Journal of Molecular and Cellular Cardiology. 2005;**39**(2):327-334

[63] Sobrevia L, Yudilevich DL, Mann GE. Activation of A2-purinoceptors by adenosine stimulates L-arginine transport (system y^+) and nitric oxide synthesis in human fetal endothelial cells. The Journal of Physiology. 1997;**499**(Pt 1):-40. DOI: 135

[64] Muñoz G, San Martín R, Farías M, Cea L, Vecchiola A, Casanello P, et al. Insulin restores glucose inhibition of adenosine transport by increasing the expression and activity of the equilibrative nucleoside transporter 2 in human umbilical vein endothelium. Journal of Cellular Physiology. 2006;**209**(3):826-835

[65] San Martín R, Sobrevia L. Gestational diabetes and the adenosine/L-arginine/nitric oxide (ALANO) pathway in human umbilical vein endothelium. Placenta. 2006;**27**(1):1-10

[66] González M, Rojas S, Avila P, Cabrera L, Villalobos R, Palma C, et al. Insulin reverses D-glucose-increased nitric oxide and reactive oxygen species generation in human umbilical vein endothelial cells. PLoS One. 2015;**10**(4):e0122398

[67] Kerr PM, Tam R, Narang D, Potts K, McMillan D, McMillan K, et al. Endothelial calcium-activated potassium channels as therapeutic targets to enhance availability of nitric oxide. Canadian Journal of Physiology and Pharmacology. 2012;**90**(6):739-752

[68] Sandow SL, Grayson TH. Limits of isolation and culture: Intact vascular endothelium and BKCa. American Journal of Physiology. Heart and Circulatory Physiology. 2009;**297**(1):H1-H7

[69] Wiecha J, Reineker K, Reitmayer M, Voisard R, Hannekum A, Mattfeldt T, et al. Modulation of Ca^{2+}-activated K^+ channels in human vascular cells by insulin and basic fibroblast growth factor. Growth Hormone & IGF Research: Official Journal of the Growth Hormone Research Society and the International IGF Research Society. 1998;**8**(2):175-181

[70] O'Malley D, Harvey J. Insulin activates native and recombinant large conductance Ca(2+)-activated potassium channels via a mitogen-activated protein kinase-dependent process. Molecular Pharmacology. 2004;**65**(6):1352-1363

[71] Rojas S, Basualto E, Valdivia L, Vallejos N, Ceballos K, Peña E, et al. The activity of IKCa and BKCa channels contributes to insulin-mediated NO synthesis and vascular tone regulation in human umbilical vein. Nitric Oxide: Biology and Chemistry. 2020;**99**:7-16

[72] Cabrera L, Saavedra A, Rojas S, Cid M, Valenzuela C, Gallegos D, et al. Insulin induces relaxation and decreases hydrogen peroxide-induced vasoconstriction in human placental

vascular bed in a mechanism mediated by calcium-activated potassium channels and L-arginine/nitric oxide pathways. Frontiers in Physiology. 2016;**7**:529

[73] Sand A, Andersson E, Fried G. Nitric oxide donors mediate vasodilation in human placental arteries partly through a direct effect on potassium channels. Placenta. 2006;**27**(2-3):181-190

Chapter 4

The Role of Physical Activity on Insulin Resistance-Associated Endothelial Dysfunction

Shruti M. Gandhi, Eric S. Nylen and Sabyasachi Sen

Abstract

Enhanced physical activity and cardiorespiratory fitness significantly impact morbidity and mortality across the spectrum of noncommunicative chronic illnesses experienced by modern lifestyles. Physical activity itself prompts an intricate interplay of physiological responses across vital organ systems including microvascular adaptations to optimize nutrient, oxygen, and hormone delivery, some of which involves insulin-mediated regulation. Insulin has been known to act on the vasculature in multiple ways by its effects on endothelium and skeletal muscle blood flow. This is important to understand as it has implications for conditions associated with insulin resistance (IR) such as obesity, metabolic syndrome, prediabetes, diabetes, and polycystic ovarian syndrome among others. These conditions are associated with increased morbidity and mortality contributed by endothelial dysfunction via increased atherosclerosis, hypertension, and increased free fatty acid levels. In this chapter, we will discuss the effects of insulin on the vasculature, IR on the endothelium, and lastly, what impact physical activity may have on such processes.

Keywords: physical activity, insulin resistance (IR), endothelium

1. Introduction

The mechanisms behind the clinical improvements following exercise and the possible roles of endothelium and adipose tissue towards tissue re-modeling and regeneration are poorly understood.

The cellular changes resulting from exercise in overweight or obese population are not fully documented. However, the incidence of overweight and obese population who are insulin resistant is gradually increasing. There seems to be an intimate relationship between fat hypertrophy, fat inflammation, and vascular supply in metabolic syndrome states such as prediabetes. The vasculature and endothelium in metabolic syndrome or subjects with prediabetes and insulin resistance are prone to ROS accumulation and inflammation. Exercise appears to improve endothelial dysfunction in insulin resistant cohort though cell-based data is lacking. The favorable impact of exercise on cardio-metabolic health depends in part on the concomitant exercise-induced reduction of adiposity and fat-based inflammation and insulin resistance.

2. Insulin as a vascular hormone

Insulin acts as a vascular hormone, mediating its action by several mechanisms including its effect on cardiac output, endothelium, type and location of vessel, and skeletal muscle [1].

Cardiac output: it has been established that insulin combined with glucose infusion causes an increase in cardiac output (CO) [2]. High concentrations of insulin in humans of 70 μU/mL cause a 15% rise in CO by increasing heart rate and stroke volume [3]. The rise in CO is associated with a decrease in mean arterial pressure and in turn, a reduction in systemic vascular resistance.

Endothelium: insulin directly acts on the vascular endothelium by binding to insulin receptors, insulin-like growth factor I (IGF-I) receptors and hybrid insulin/IGF-I receptors [4]. With the binding of insulin to these endothelial receptors, both vasodilator (i.e., nitric oxide, NO) and vasoconstrictor (i.e., endothelin 1, ET-1) substances are released to balance vascular tone. NO causes vasodilation of the vessels via the activation of insulin receptor substrate-1 (IRS-1) leading to phosphatidylinositol 3-kinase (PI-3 kinase)/protein kinase B (Akt.) phosphorylation of

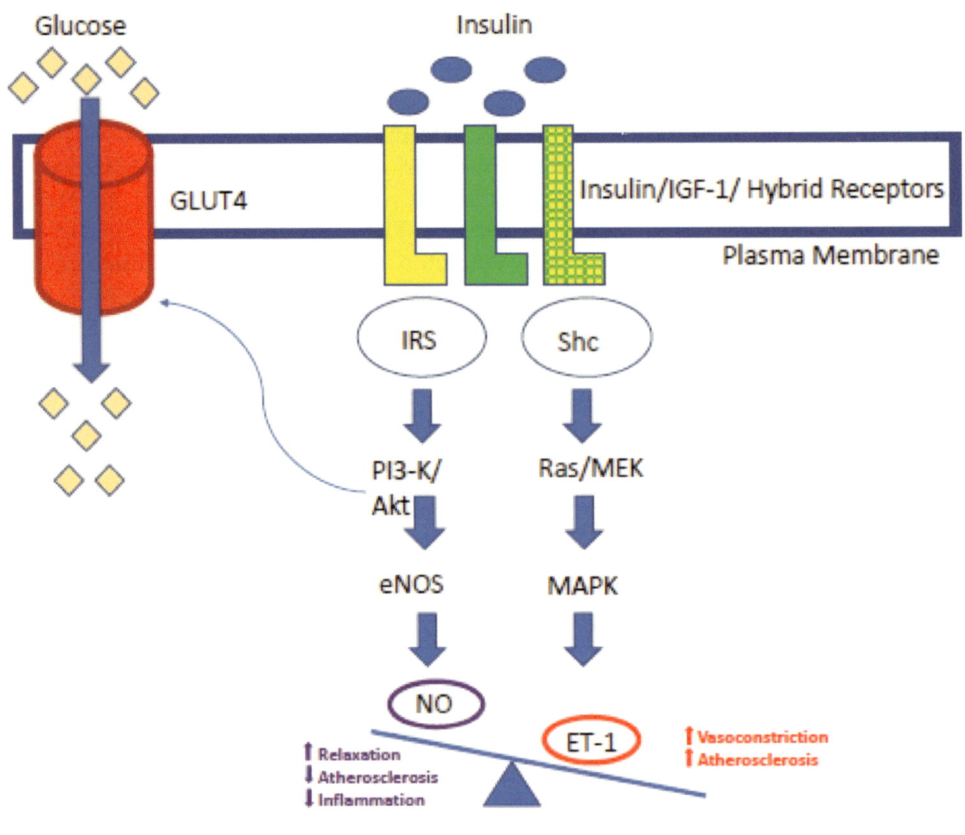

Figure 1.
Pathway of insulin-mediated release of nitric oxide (NO) and endothelin-1 (ET-1), two peptides that influence vasomotor tone and risk of atherosclerosis. Insulin directly affects the vascular endothelium by binding to receptors present on the endothelium which expresses not only insulin receptors but also insulin-like growth factor I (IGF-I) receptors and hybrid insulin/IGF-I receptors. These hybrid receptors are expressed 5- to 10-fold higher concentration than insulin receptors. Potentially, an over-abundance of IGF-1 receptor may result in vascular insulin resistance by promoting hybrid receptor formation which does not respond to physiological insulin concentrations while decreasing insulin receptor availability. As described in the text, insulin initiates an intracellular cascade of steps resulting in NO or ET-1. GLUT-4 transports glucose across the cellular membrane following insulin activation or muscle contraction via a complex of intermediary substrates. The former is initiated by the PI3K pathway, while muscle contraction by a PI3K independent pathway.

Vessel	Role	Insulin action	Mechanism
Conduit arteries	Regulate arterial compliance, blood pressure	Increases relaxation, increases compliance	Likely NO
Resistance arterioles	Determines vascular resistance	Dilation, decrease vascular resistance	NO
Terminal arterioles, capillaries	Regulate insulin delivery to muscle. Also exchange nutrients, oxygen and hormones with muscle	Muscle glucose uptake, recruit muscle microvasculature	Insulin, muscle contraction, angiotensin GLP1, adiponectin
Skeletal muscle	Muscle contraction	Increase blood flow, increase glucose uptake	NO, translocation of GLUT4 receptors

Table 1.
Role of insulin on various vessels and mechanism.

endothelial NO synthase (eNOS) whereas the ET-1 signaling pathway involves the mitogen activated protein kinase (MAPK) (**Figure 1**). During states of high insulin concentrations, as seen in euglycemic hyperinsulinemic clamp studies or postprandially, insulin's vasodilatory effect through NO predominates. NO not only decreases vascular tone but also decreases vascular smooth muscle cell (VSMC) proliferation and reduces binding of inflammatory cells and platelet aggregation [5].

Type and location of vessels: insulin's action on the vasculature varies depending on its site of action along the arterial tree which can include the conduit arteries, the resistance arterioles, precapillary arterioles and the capillaries (**Table 1**). It also acts on the skeletal muscle vasculature and can have effects locally.

The conduit arteries are large arteries which regulate arterial compliance in response to the ejection volume and stretch in order to maintain blood pressure. Insulin increases compliance by vasodilation of these vessels in response to NO release. In human studies with insulin infusion, the responsiveness of the femoral artery to methacholine-induced vasodilation is increased [4].

The resistance arterioles regulate blood pressure and total blood flow to tissues. They determine vascular resistance as a change in the size of the vessel (lumen size) can significantly increase or decrease resistance and thus the amount of blood supply to the tissues. Insulin via NO production causes dilation of these vessels, decreases resistance, and increases blood flow.

The microvasculature including the terminal arterioles, capillary networks and venules regulate insulin delivery to muscle tissues. Insulin action here promotes glucose uptake, recruitment of muscle vasculature and its own trans-endothelial transport [4]. This allows for exchange of nutrients, oxygen, and hormones to the muscle with removal of metabolic waste.

Skeletal muscle: during exercise, skeletal muscle blood flow, capillary recruitment, and GLUT4 translocation to the sarcolemma and T-tubules are augmented which is essential for glucose uptake and oxidation. Insulin targets the skeletal muscle by increasing blood flow and glucose uptake, the latter mediated by translocation of GLUT4 transporters to the sarcolemma and transverse tubules as well as to the surface of the cell (independent of insulin).

3. Insulin resistance (IR) and vasculature

Insulin resistance is characterized by a state of compensatory hyperinsulinemia due to changes in insulin secretion and/or insulin clearance [6] leading to mild forms of glucose intolerance, dyslipidemia (high triglycerides, low HDL, small dense LDL), and hypertension termed the "insulin resistance syndrome". As discussed, during physiological conditions, insulin binding to endothelial receptors leads to phosphorylation of downstream substrates including activation of the IRS-1, PI3K pathway and subsequent recruitment of GLUT4 to mediate glucose transport into muscle and other tissues [7]. However, In the IR state, the IRS-1-PI3K-Akt-NO pathway is muted while the MAPK pathway remains intact [8]. The unopposed action of ET-1 leads to a shift towards vasoconstriction, increased arterial stiffness, hypertension, and tissue hypoxia. In addition to this decrease in NO bioavailability, increases in oxidative stress, inflammatory markers, and pro-thrombotic mediators (i.e., increased plasma von Willebrand factor, decreased lipoprotein lipase activity) are seen. More direct evidence for endothelial IR was shown in freshly isolated arterial endothelial cells where the insulin-induced eNOS-phosphorylation was negatively associated with oxidative stress markers [9].

During states of IR each vascular site becomes affected and contributes to an increase in atherosclerosis. At the level of the conduit arteries, IR causes decreased compliance with a concomitant increase in vessel stiffness, which is a predictor of coronary artery disease and stroke [10]. The impaired vasodilatory action of insulin on the resistance arterioles leads to decrease in blood flow to the tissues it supplies. For example, Baron et al. demonstrated that the inhibition of NO production (similar to states of IR) causes a decrease in blood flow and glucose uptake in the leg. The terminal arterioles in patients with IR showed a blunted response to mixed meal in brachial blood flow and forearm microvascular recruitment compared to lean subjects [3].

Oxidative stress: hyperglycemia due to IR also induces generation of reactive oxygen species (ROS) by activation of the NADPH oxidase system. ROS activates multiple pathways linked with cell growth, proliferation and modifies NO bioavailability. One such pathway includes the renin-angiotensin system which is inappropriately activated in settings of IR. Interestingly, during continuous insulin infusion, Angiotensin 2 receptor antagonism resulted in whole-body insulin resistance and attenuation of microvasculature recruitment [5]. The mechanism may involve increased binding to Angiotensin 1 receptors, which have been shown to increases oxidative stress and cause vasoconstriction through decreased bioavailability of *e*NOS and increased ROS. Chai et al. has shown that when AT2R is blocked, there is decreased microvascular blood flow by 80% along with reduced glucose extraction [11].

4. The effect of exercise on insulin resistance and vasculature

It is well established that exercise augments insulin signaling independent of PI3K, while the combination of skeletal muscle contraction and insulin additively enhances glucose transport via GLUT4 translocation. A plethora of studies have reported that regular physical activity is effective in patients with IR, such as type 2 diabetes, prediabetes and metabolic syndrome, in improving glucose tolerance, insulin sensitivity, glycosylated hemoglobin levels (HbA1c) and morbidity and mortality [12]. For instance, adults with IR were found to have improvements in hepatic and peripheral insulin sensitivity after 12 weeks of aerobic exercise. Shorter term studies (i.e., 7 days) have also demonstrated similar improvements in insulin sensitivity in obese patients [13]. Lifestyle interventions such as diet modifications added to 12 weeks exercise training showed further enhancements in addition to insulin sensitivity including fatty acid oxidation, post-prandial hyperinsulinemia and systolic resting blood pressure [14].

Exercise and the endothelium: exercise causes several adaptations to IR in the vasculature in both the skeletal muscle and endothelium. Vessel wall shear stress generated by exercise activates the PI3k/Akt/NO signaling pathway resulting in increased expression of eNOS and improved endothelial vasodilation and vascular remodeling [15]. Vessels with high shear stress are considered anti-atherogenic (low ET-1, high NO bioavailability) as opposed to low shear stress environments (high ET-1, low NO bioavailability). In patients with Type 2 diabetes, 8 weeks of combined aerobic and resistance exercises improved flow mediated dilation (FMD) of the brachial artery suggesting increased shear stress and improved endothelial vasodilation [16]. Exercise improved FMD, microvascular perfusion in muscles of older adults relative to sedentary adults in nondiabetic subjects with metabolic syndrome [14]. Finally, insulin sensitization without exercise also augments FMD in prediabetes [17].

Skeletal muscle: during exercise, blood flow to the skeletal muscle increases up to 100-fold through vasodilation and recruitment of capillaries to help maximize oxygen extraction as well as insulin delivery to the skeletal muscle [5]. Pivotal studies investigated the vascular effects of exercise training on insulin [7, 15, 18]. Single leg cycle exercises over a 10-week period improved insulin stimulated glucose uptake and vasodilation in the trained limb post exercise training for both healthy and IR subjects. Moreover, insulin stimulated vasodilation in the lower limb is greater in endurance trained athletes compared to otherwise healthy sedentary controls [7].

5. Endothelium and endothelial progenitor cells

Endothelial cells constitute the innermost layer of blood vessel and promote vascular homeostasis and angiogenesis. Endothelial cells can secrete several mediators that can alternatively mediate vasoconstrictors, such as endothelin-1 and thromboxane A2, or vasodilators, such as NO, prostacyclin, and endothelium-derived hyperpolarizing factor. Since hyperglycemia and IR can negatively impact NO secretion from the endothelium, with vasoconstriction, vessel wall stiffness, platelet aggregation and diminished angiogenesis, there is a counter-regulatory reparatory cellular response by circulating endothelial progenitor cells (EPCs). These are immature bone marrow derived cells that can differentiate into mature endothelial cells. These cells home in on areas that experience vascular injury or ischemia by way of circulating growth factors and cytokines to initiate repair of the endothelial surface and stimulate neovascularization and angiogenesis. In conditions such as diabetes with vascular damage, the presence of diminished circulating EPCs constitute cellular biomarkers of compromised cardiovascular health [19]. In subjects with IR such as metabolic syndrome, decreased EPC number and impaired functionality prognosticates increased cardiovascular risk [20]. Interestingly, exercise promotes the production and numbers of EPCs [19, 21] putatively related to the anti-apoptotic effect of NO [22]. EPCs are also stimulated by exercise in aging studies. Moreover, the migratory function of EPCs is improved by exercise in subjects with IR [19, 23]. The degree of exercise dose appears to influence the overall EPC response [23]. In presence of insulin resistance but not overt diabetes, CPAP therapy improves endothelial health and EPC parameters [24].

Exercise, endothelium and fat derived mesenchymal stromal cells (MSCs): clinical trials are necessary to investigate the possible cellular and molecular pathways that may impact endothelium and fat metabolism. Identification of the pathways that influence crosstalk between endothelium and fat, and thereby improve cardio-metabolic health in the elderly and young subjects is important to identify. The process will help to identify genes and cell differentiation pathways that may change fat derived stem cell differentiation, following exercise training in the elderly and the young subject cohorts, which will subsequently influence the mesenchymal structures of our body. Our study [25] appears to indicate that exercise promotes osteogenic differentiation but not myogenic differentiation in the middle-aged veteran population with mean age of 51 years. However, whether osteogenic differentiation of adipose tissue derived mesenchymal stromal cells (MSCs) also occurs in young and the elderly is unknown. Myogenic differentiation in response to exercise is well documented [26], and different types of exercise appear to influence the differentiation depending on plasma-based differentiation factors [26]. However, the exact mechanism of how exercise modifies mesenchymal stromal cell (MSC) differentiation in the body needs further investigation. Prior to our recent studies, we would have hypothesized that exercise will promote myogenic differentiation in all age groups, however the differentiation of stem cells may be dependent upon the need of the body to regenerate a particular tissue lineage at a particular age. For example, exercise promotes myogenic differentiation in the young [26] whereas endothelial function improvement and bone regeneration may be more important in the elderly [27, 28].

6. Summary

Exercise is an important modifiable risk factor that significantly attenuates cardiovascular morbidity and mortality. Physical activity is associated with enhanced cardiorespiratory fitness which significantly attenuates IR, and some of these effects are mediated by augmented endothelial action of insulin. These vascular effects of exercise include an increase in endothelium-dependent vasodilation through increased NO bioavailability, suppression of ET-1, increased capillary density, and reduction in ROS.

Finally, exercise appears to rejuvenate endothelial function by recruitment of exercise responsive EPCs and influences MSC differentiation.

Disclosures

None.

Author details

Shruti M. Gandhi[1], Eric S. Nylen[1] and Sabyasachi Sen[2]*

1 VAMC – Veteran's Affairs Medical Center, George Washington University School of Medicine, Washington, DC, USA

2 Department of Medicine, The George Washington University, Washington, DC, USA

*Address all correspondence to: ssen1@gwu.edu

References

[1] Kokkinos P, Narayan P, editors. Cardiorespiratory Fitness in Cardiometabolic Diseases. Springer; 2019. Available from: https://link.springer.com/book/10.1007/978-3-030-04816-7

[2] Baron AD, Steinberg HO. Vascular actions of insulin in health and disease. In: Sowers JR, editor. Endocrinology of the Vasculature Contemporary Endocrinology. 1996. pp. 96-104. Chapter 7. Available from: https://link.springer.com/chapter/10.1385/1-59259-908-7:001

[3] Baron AD, Brechtel G. Insulin differentially regulates systemic and skeletal muscle vascular resistance. The American Journal of Physiology. 1993;**265**:E61-E67

[4] Zheng C, Liu Z. Vascular function, insulin action and exercise: An intricate interplay. Trends in Endocrinology and Metabolism. 2015;**26**(6):297-304

[5] Manrique C, Lastra G, Sowers JR. New insights into insulin action and resistance in the vasculature. Annals of the New York Academy of Sciences. 2014;**1311**(1):138-150

[6] Jones CN, Pei D, Staris P, Polonsky KS, Chen YD, Reaven GM. Alterations in the glucose-stimulated insulin secretory dose-response curve and in insulin clearance in nondiabetic insulin-resistant individuals. The Journal of Clinical Endocrinology and Metabolism. 1997;**82**:1834-1838

[7] Hardin DS, Azzarelli B, Edwards J, Wigglesworth J, Maianu L, Brechtel G, et al. Mechanisms of enhanced insulin sensitivity in endurance-trained athletes: Effects on blood flow and differential expression of GLUT 4 in skeletal muscles. Journal of Clinical Endocrinology and Metabolism. 1995;**80**:2437-2446

[8] Muniyappa R, Sowers JR. Role of insulin resistance in endothelial dysfunction. Reviews in Endocrine and Metabolic Disorders. 2013;**14**(1):5-12

[9] Masaki N, Idop Y, Yamada T, et al. Endothelium insulin resistance of freshly isolated arterial endothelial cells from radial sheaths in patients with suspected coronary artery disease. Journal of the American Heart Association. 2019;**8**:e010816

[10] Mattace-Raso FU, van der Cammen TJ, Hofman A, van Popele NM, Bos ML, Schalekamp MA, et al. Arterial stiffness and risk of coronary heart disease and stroke: The Rotterdam study. Circulation. 2006;**113**:657-663

[11] Chai W, Wang W, Dong Z, Cao W, Liu Z. Angiotensin II receptors modulate muscle microvascular and metabolic responses to insulin in vivo. Diabetes. 2011;**60**(11):2939-2946

[12] Nylen ES, Kokkinos P. Physical activity, cardiorespiratory fitness, and the diabetes spectrum. In: Kokkinos P, Narayan P, editors. Cardiorespiratory Fitness in Cardiometabolic Diseases. Springer; 2019. Available from: https://link.springer.com/book/10.1007/978-3-030-04816-7

[13] Malin SK, Haus JM, Solomon TP, Blaszczak A, Kashyap SR, Kirwan JP, et al. Insulin sensitivity and metabolic flexibility following exercise training among different obese insulin-resistant phenotypes. American Journal of Physiology. Endocrinology and Metabolism. 2013;**305**:E1292-E1298

[14] Seligman BG, Polanczyk CA, Santos AS, Foppa M, Junges M, Bonzanini L, et al. Intensive practical lifestyle intervention improves endothelial function in metabolic syndrome independent of weight

loss: A randomized controlled trial. Metabolism. 2011;**60**:1736-1740

[15] Olver DT, Laughlin HM, Padilla J. Exercise and vascular insulin sensitivity in the skeletal muscle and brain. Exercise and Sports Sciences Reviews. 2019;**47**(2):66-74

[16] Maiorana A, O'Driscoll G, Cheetham C, Dembo L, Stanton K, Goodman C, et al. The effect of combined aerobic and resistance exercise training on vascular function in type 2 diabetes. Journal of the American College of Cardiology. 2001;**34**:860-866

[17] Avena R, Mitchell ME, Nylen ES, Curry KM, Sidawy AN. Insulin action enhancement normalizes brachial artery vasoactivity in patients with peripheral vascular disease and occult diabetes. Journal of Vascular Surgery. 1998;**28**(6):1024-1031

[18] Dela F, Larsen JJ, Mikines KJ, Ploug T, Petersen LN, Galbo H. Insulin stimulated muscle glucose clearance in patients with NIDDM. Effects of one-legged physical training. Diabetes. 1995;**44**:1010-1020

[19] Sen S, Witkowski S, Lagoy A, Islam AM. A six-week home exercise program improves endothelial function and CD34+ circulating progenitor cells in patients with prediabetes. Journal of Endocrinology & Metabolism. 2015;5:163-171

[20] Jialal I, Devaraj S, Singh U, Huet BA. Decreased number and impaired functionality of endothelial progenitor cells in subjects with metabolic syndrome: Implications for increased cardiovascular risk. Atherosclerosis. 2010;**211**(1):297-302

[21] Laufs U, Werner N, Link A, Endres M, Wassmann S, Jurgens K. Physical training increases endothelial progenitor cells, inhibits neointima formation, and enhances angiogenesis. Circulation. 2004;**109**:220-226

[22] Jenkins NT, Witowski S, Spangenburg EE, Hagberg JM. Effects of chronic and acute endurance exercise in intracellular nitric oxide in putative endothelial progenitor cells: Role of NADPH oxidase. American Journal of Physiology-Heart and Circulatory Physiology. 2009;**297**:H1798-H1805

[23] Kundu N, Domingues CC, Sen S. Effect of exercise on adult stem cells. In: Kokkinos P, Narayan P, editors. Cardiorespiratory Fitness in Cardiometabolic Diseases. Springer; December 2018. DOI: 10.1007/978-3-030-04816-7_4

[24] Domingues CC, Dore FJ, Cho A, Ahmadi N, Kropotova Y, Kundu N, et al. Reassessing the effects of continuous positive airway pressure (CPAP) on arterial stiffness and peripheral blood derived CD34+ progenitor cells in subjects with sleep apnea. Stem Cell Research & Therapy. 2019;**10**(1):147. DOI: 10.1186/s13287-019-1251-8

[25] Kundu N, Domingues CC, Nylen ES, Paal E, Kokkinos P, Sen S. Endothelium-derived factors influence differentiation of fat-derived stromal cells post-exercise in subjects with prediabetes. Metabolic Syndrome and Related Disorders. 2019. DOI: 10.1089/met.2018.0121

[26] Vitucci D, Imperlini E, Arcone R, Alfieri A, Canciello A, Russomando L, et al. Serum from differently exercised subjects induces myogenic differentiation in LHCN-M2 human myoblasts. Journal of Sports Sciences. 2018;**36**(14):1630-1639

[27] Witkowski S, Lockard MM, Jenkins NT, Obisesan TO, Spangenburg EE, Hagberg JM. Relationship between circulating progenitor cells, vascular function and oxidative stress with long-term training and short-term detraining in older men.

Clinical Science (London, England). Feb. 2010;**118**(4):303-311

[28] Colberg SR, Sigal RJ, Fernhall B, Regensteiner JG, Blissmer BJ, Rubin RR, et al. Exercise and type 2 diabetes: The American College of Sports Medicine and the American Diabetes Association: Joint position statement. Diabetes Care. 2010;**33**(12):e147-e167

Chapter 5

Therapeutic Applications and Mechanisms of YC-1: A Soluble Guanylate Cyclase Stimulator

Chieh-Hsi Wu, Chun-Hsu Pan and Ming-Jyh Sheu

Abstract

Nitric oxide (NO) is an essential endogenous vasodilator to maintain vascular homeostasis, whose effects are mainly mediated by NO-dependent soluble guanylate cyclase (sGC) which catalyzes the synthesis of cyclic guanosine monophosphate (cGMP), a critical mediator of vascular relaxation. YC-1, a novel NO-independent sGC stimulator, was first introduced as an inhibitor of platelet aggregation and thrombosis. Accumulating studies revealed that YC-1 has multiple medication potentials to use for a broad spectrum of diseases ranging from cardiovascular diseases to cancers. In contrast to NO donors, YC-1 has a more favorable safety profile and low medication tolerance. In this chapter, we introduce canonical and pathological roles of NO, review activations, and regulatory mechanisms of YC-1 on NO-independent sGC/cGMP pathway and present the potential pharmacological applications and molecular mechanisms of YC-1.

Keywords: nitric oxide, soluble guanylate cyclase, YC-1

1. Introduction

Since the discoveries of the biological effects of NO on physiological actions mediated by cGMP, delineation of the molecular mechanism of NO actions and understanding of NO activation of guanylate cyclase (GC) and the subsequent signal processes have been greatly advanced [1]. NO can function as an intracellular messenger, an autacoid, a paracrine substance, a neurotransmitter, or as a hormone that can be carried to distant sites for effects [1, 2]. It is therefore a unique simple molecule with diversified physiological functions.

2. Canonical function of NO

NO, initially known as the endothelium-derived relaxing factor (EDRF), is a gas molecule and free radical with an unpaired electron which has been shown to be involved in an ever-growing list of biological processes. NO generated in the tissue binds to major physiological target, haem moiety of GC, activating the cGMP cascade. The GC family is composed of two members including membrane-bound GC and soluble GC (sGC). Membrane-bound GC is a receptor responsive to atrial natriuretic peptide (ANP), and sGC acts as the NO sensor. NO exerts its biological

IntechOpen

effects by activating sGC to increase the cGMP level and vascular effects known to be mediated by cGMP such as vasodilation, inhibition of platelet aggregation, and inflammatory reaction. Cyclic GMP modulates a number of signaling processes downstream of NO. The NO-cGMP cascade can be regulated by pharmacological modulation of protein kinases, phosphodiesterases (PDE), and ion channels to alter vascular tones as well as endothelial and vascular smooth muscle cell growth. Pharmacological alteration of the NO level has been a major strategy to develop therapeutic agents for cardiovascular diseases.

Deguchi and his colleagues found that GC activity in the supernatant of neuroblastoma and brain preparations were activated by L-arginine which has been identified as an endogenous activator of sGC [3]. Hibbs et al. noted the latter that the cytotoxic properties of macrophages in co-cultures with tumor cells could be enhanced with L-arginine but suppressed by N-N-methyl-arginine (LNMA), an inhibitor of nitric oxide synthase (NOS) [4]. This cytotoxicity action was accompanied by accumulation of nitrite in the conditioned medium. These important studies provide the insight to identify a pathway of L-arginine metabolism that could produce NO and nitrite.

NOS is a group of isozymes which convert L-arginine to L-hydroxyarginine and subsequently to NO and L-citrulline through cofactors including reduced nicotinamide-adenine dinucleotide phosphate (NADPH), flavin adenine dinucleotide (FAD), flavin mononucleotide (FMN), and tetrahydrobiopterin (BH4) [5]. The first NOS isoform to be identified was the neuronal NOS (nNOS or named as NOS-1) [6]. This was followed shortly thereafter by inducible NOS (iNOS), also known as type II NOS (NOS-2) [7], and then by endothelial NOS (eNOS or named as NOS-3) [8]. NOS can also be inactivated by asymmetric dimethyl arginine (ADMA), an endogenous and competitive inhibitor of NOS [9–11].

3. Pathological role of NO

NO is essential in the maintenance of vascular homeostasis including smooth muscle relaxation, inhibition of platelet aggregation, attenuation of vascular smooth muscle cell (VSMC) proliferation, neurotransmission, and immune defense [12]. Therefore, the impaired NO pathway has been implicated in endothelial dysfunction and pathogenesis of a number of diseases featuring inflammatory reaction. These include arthritis, myocarditis, colitis, and nephritis. Altered NO synthesis has been noted in selected pathologic conditions such as amyotrophic lateral sclerosis, cancer, diabetes, and neurodegenerative diseases [13, 14]. In general, physiological NO actions on target tissues are brief, reversible, and dependent on the downstream cGMP-dependent signaling events. Conversely, the pathological actions noted with excessively and sustained NO production involved NO interaction with superoxide to generate peroxynitrite, a highly reactive free radical which exhibits the toxic actions of potent oxidants. Peroxynitrite, independent of the cGMP signaling events, has been implicated in oxidative injury noted in a number of disease models [15–17]. In addition to its free radical actions, peroxynitrite inactivates prostacyclin synthase to reduce prostacyclin levels, leading to vascular dysfunction [18].

ADMA, a risk factor for cardiovascular diseases, inhibits NOS to reduce biosynthesis of NO, resulting in impaired blood flow, accelerated atherogenesis, and suppressed angiogenesis [19]. ADMA is involved in the development of endothelial dysfunction. In essential hypertension patients, the L-arginine and ADMA levels are elevated and inversely related to endothelial function [20, 21]. Endothelial function depends on the integrity of eNOS and the availability and vascular signaling of

NO. In clinical settings, endothelial dysfunction is important because it may develop hypertension and atherosclerosis and therefore is a predictor in ensuing cardiovascular diseases [22]. In hyperhomocysteinemia, an increase in ADMA has been linked to impaired vascular endothelial function. Elevated homocysteine levels exert inhibitory effects on the expression or activation of dimethylarginine dimethylaminohydrolase (DDAH) [23–27]. Two isoforms of DDAH, DDAH-1 and DDAH-2, were identified in tissues expressing nNOS and eNOS, respectively [28]. Both DDAH isoforms are expressed widely in different organs, with higher content found in the liver and kidney [29]. Similarly, endothelial dysfunction has also been found in hypercholesterolemia. Several studies indicated that hypercholesterolemia may cause a decline in DDAH activity and an increase in the ADMA level [30, 31]. Böger et al. also found that exposure of cultured endothelial cells to oxidized low-density lipoprotein (oxLDL) cholesterol resulted in ADMA accumulation in the culture medium [31]. Oxidized LDL could cause endothelial dysfunction in complex mechanisms including reduction of eNOS expression [31], to trigger endothelial apoptosis [32] and to inhibit vascular endothelial growth factor (VEGF)-induced endothelial proliferation [33]. Furthermore, oxLDL impairs NO-induced stimulation of cGMP accumulation [34]. Patients with cardiac syndrome X (CSX) have higher levels of ADMA and increased mean common carotid intima-media thickness that are ascribed to ADMA effects on NO bioavailability resulting in endothelial dysfunction and subsequently impede microvascular circulation, which are the leading mechanisms in the development of CSX [10, 35, 36]. ADMA also plays important roles in endothelial dysfunction in subjects with chronic kidney failure [9, 37, 38]. ADMA is metabolized to L-citrulline *via* the action of DDAH-1, which is highly expressed in the kidney [29]. There is a strong association between impairment of renal function and elevation of ADMA content [9, 39]. Microangiopathy-related cerebral damage (MARCD) is a cerebrovascular disease caused by arteriosclerosis in deep white matter, which includes lacunar infarction and white matter hyperintensity [40]. Arteriosclerosis in deep white matter resulting from acute and chronic ischemia is probably responsible for the development of MARCD [41]. Several potential risk factors for arteriosclerosis have been evaluated in patients with MARCD [42, 43]. NO is involved not only in regulating cerebral blood flow but also in preventing arteriosclerosis by inhibiting fibrosis and proliferation of smooth muscle cells in the arterial wall [44]. In fact, NOS inhibitors and functional single-nucleotide polymorphisms in the eNOS gene have been shown to be correlated with MARCD [44, 45]. Excessive NO production could also be a problem in the progression of the disease such as glaucoma. Increased NO generated by iNOS in astrocytes and microglia in the optic nerve head of patients with glaucoma may contribute to the optic neuropathy associated with this disease. The pharmacological use of an inhibitor of iNOS, aminoguanidine, significantly prevents the loss of retinal ganglion cells [46].

4. Novel compounds for NO-independent sGC/cGMP activation

Organic NO donors such as nitrite and nitroglycerin are successful examples in clinical practice for more than a century. However, formation of harmful intermediate, peroxynitrite, and the long-term treatment with NO donors resulting in drug resistance limit the clinical applications of NO donor compounds. To overcome these obstacles, the novel agents for triggering sGC/cGMP cascade in NO-independent manner have been developed.

A series of 1-(substituted benzyl)-3-(substituted aryl)-condensed pyrazole derivatives were synthesized and identified as class novel antiplatelet agents [47, 48]. As one of the most promising analogues, 1-benzyl-3-(5′-hydroxymethyl-2′-furyl) indazole (YC-1) was selected for further investigation. The physiological property of YC-1 in stimulation of sGC was demonstrated by Ko and colleagues [49]. Potential regulatory mechanisms of YC-1 on cardiovascular protections were summarized in **Figure 1**. Ko et al. showed that YC-1 is an antithrombotic agent. It inhibits platelet aggregation by increasing platelet cGMP levels in an NO-independent manner. YC-1 action was noted to exert its antiplatelet effect through the activation of NO-independent sGC/cGMP pathway [50]. Nearly, all the newer generations of sGC stimulator except acryl-acrylamide family have been derived based on YC-1 as the parent compound [51]. YC-1 and its successors all require the presence of a reduced haem moiety within sGC to stimulate sGC, but they also act in synergy with NO by binding NO or iron-free precursor of haem to structurally resemble the NO-haem complex and stabilize sGC in its active configuration [52–54]. Stasch et al. also reported that YC-1 and its derivate, BAY 41-2272, bind to regulatory sites (cys 238 and cyst 243 regions) in the α1-subunit of sGC, resulting in conformational change and subsequent activation of recombinant sGC by NO-independent but haem-dependent mechanism [55]. Mulsch et al. also noted that the combined effect of nitrovasodilators and YC-1 in cultured VSMCs and isolated rabbit aortic rings reflected the direct synergistic action of YC-1 and NO on the sGC [56]. Wohlfart et al. reported that YC-1 can stimulate synthesis and release NO in endothelial cells independent of raising the cGMP content in a calcium-dependent manner [57]. In addition, YC-1 inhibits the cGMP-specific phosphodiesterase type 5 (PDE-5)

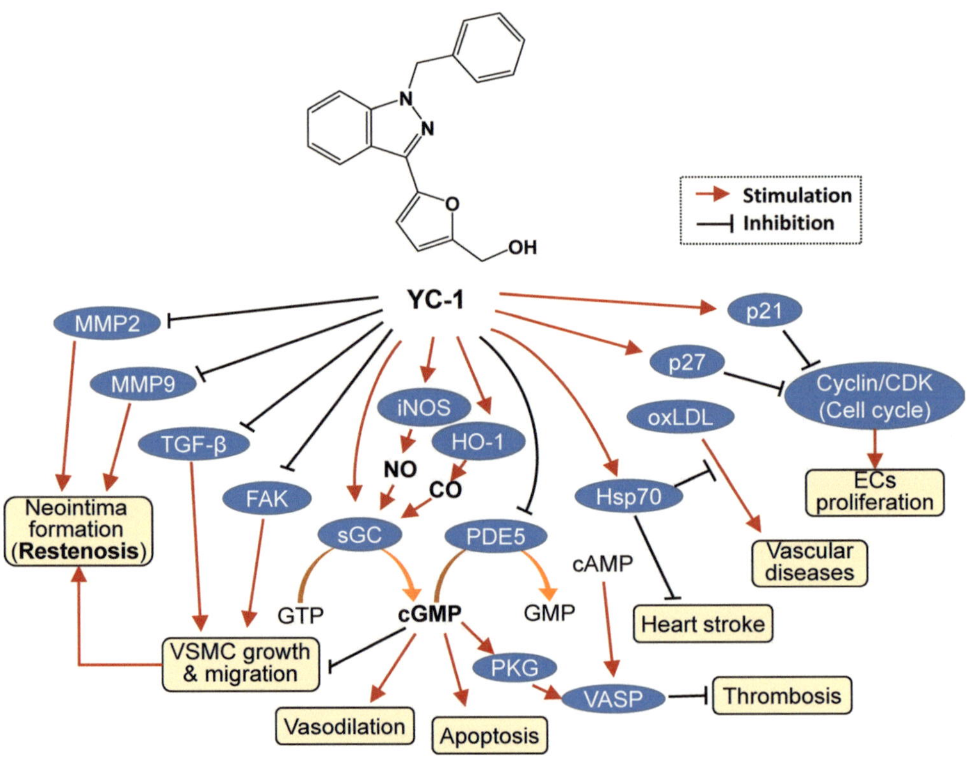

Figure 1.
Schematic overview of regulatory mechanisms of YC-1 on cardiovascular protections. cAMP, cyclic adenosine monophosphate; cGMP, cyclic guanosine monophosphate; ECs, endothelial cells; FAK, focal adhesion kinase; GMP, guanosine monophosphate; GTP, guanosine triphosphate; HO-1, heme oxygenase-1; Hsp70, heat shock protein 70; iNOS, inducible nitric oxide synthase; MMP2, matrix metalloproteinase-2; MMP-9, matrix metalloproteinase-9; oxLDL, oxidized low-density lipoprotein; PDE5, phosphodiesterase 5; PKG, protein kinase G; sGC, soluble guanylyl cyclase; TGF-β, transforming growth factor-beta; VASP, vasodilator-stimulated phosphoprotein; VSMC, vascular smooth muscle cells.

in platelets and in aortic extracts to raise cGMP levels and prolong its duration of action [58, 59]. The vasodilator-stimulated phosphoprotein (VASP) has been reported to be involved in cGMP- and cAMP-mediated antiplatelet actions [60]. Becker et al. noted that VASP is the target of YC-1 since VASP phosphorylation can be directly increased through stimulation of the cGMP/protein kinase G/VASP pathway [61].

5. Additional pharmacological activities of YC-1

In addition to the effects in antiplatelet aggregation and antithrombosis, YC-1 has been demonstrated to provide several beneficial effects including cardiovascular protections; antitumor, neuroprotective, and anti-inflammatory effects; as well as optical protections.

5.1 Cardiovascular protections

YC-1 inhibits VSMC proliferation, similar to specific guanylate cyclase inhibitors, suggesting that the antiproliferative effect of YC-1 is mediated by cGMP [62]. A similar conclusion has also been drawn by other investigators [63, 64]. As shown in **Figure 1**, NO-/cGMP-dependent processes have been suggested to modulate VSMC phenotype and the arterial response to endovascular injury [65, 66]. It has been reported that YC-1 can upregulate expression of iNOS and inducible heme oxygenase-1 (HO-1) at the transcriptional and translational level as well as stimulate sGC and cGMP production in the balloon-injured artery [63]. These results support the proposal that YC-1 can be developed as a potent new therapeutic agent for reducing restenosis *via* endogenous carbon monoxide (CO)- and/or NO-mediated cGMP-dependent processes. Wu et al. found that two important modulators, transforming growth factor (TGF)-β1 and focal adhesion kinase (FAK), responsible for VSMC proliferation and migration were reduced in content in the cultured VSMC treated with YC-1. The effect of YC-1 on preventing balloon injury-induced vascular stenosis has also been demonstrated in a rat carotid angioplasty model [64]. Liu et al. also found that YC-1 can inhibit neointima formation in balloon-injured rat carotid through suppressing the expression and actions of matrix metalloproteinase (MMP)-2 and MMP-9 [67]. YC-1 can also prevent oxLDL-mediated apoptosis by inducing heat shock protein 70 (Hsp70) expression in VSMCs suggesting its cytoprotective effect in vascular diseases [68]. Similarly, Hsp70 overexpression has also been involved in protective effect of YC-1 on heat stroke [69]. In vivo evidence shows that YC-1 and zaprinast, an inhibitor of cGMP-selective PDE, inhibit injury-induced vascular remodeling through anti-mitogenic and pro-apoptotic actions in a rat carotid artery balloon injury model [70]. Moreover, YC-1 has also been found to induce cell cycle arrest of HUVEC through upregulation of p21 and p27 protein *via* inhibition of the cyclin/cyclin-dependent kinase (CDK) system. This finding suggests that YC-1-induced antiproliferation effect in HUVEC is *via* a cGMP-independent manner [71]. Besides, the prevention effects of YC-1 on the development of hypoxia-induced pulmonary arterial hypertension (PAH), right ventricular hypertrophy (RVH), and pulmonary vascular remodeling has been clearly mentioned in animal model [72].

5.2 Antitumor effects

A growing body of evidence indicates that hypoxia-inducible factor-1 (HIF-1) contributes to tumor progression and metastasis. YC-1 inhibits HIF-1-mediated hypoxic responses [73–76]. YC-1 enhanced radiation sensitivity by inhibiting HIF-1α expression [77]. Lau et al. also found that YC-1 suppressed both synthesis and stability of HIF-1α, *via* regulation of murine double minute (Mdm2) protein [78]. In hypoxic gastric carcinoma cell and xenograft models, low-dose YC-1 combined with glucose and insulin can effectively inhibit anaerobic glycolysis and induce hypoxia-dependent apoptosis by suppressing HIF-1α expression [79].

YC-1 also enhanced chemosensitivity of hepatocellular carcinoma cells to cisplatin through a Stat3-dependent manner [80]. Similarly, YC-1 also enhanced camptothecin toxicity by activating the caspase-8, the Bid pathway, and the mitochondria-mediated apoptotic pathway in ovarian carcinoma cell lines [81]. Additionally, it has also been found that YC-1 can suppress constitutive NF-κB activation and induce apoptosis in human prostate cancer cells [82]. YC-1 inhibited VEGF- and basic fibroblast growth factor (bFGF)-mediated ERK1/ERK2 mitogen-activated protein kinase (MAPK), AKT, and protein kinase Cα (PKCα) pathways in vitro and angiogenesis in in vivo models [83]. YC-1 arrested the cell cycle in G_0/G_1 in human hepatocellular carcinoma cells by upregulating $p21^{CIP1/WAP1}$ and $p27^{KIP1}$ expression [84]. YC-1 arrested the cell cycle at S-phase and induced apoptosis by activating checkpoint kinases in several cancer cells [85]. Similarly, YC-1 can also increase p21 protein and decrease cyclins and CDKs to induce G0/G1 phase arrest as well as activate caspases and disrupt the mitochondrial membrane potential to trigger mitochondria-dependent apoptosis in cisplatin-resistant human oral cancer CAR cells [86]. Additionally, apoptotic mechanism of YC-1 may also be mediated by activating JNK phosphorylation and upregulating FasL and Fas receptor clustering to activate caspase-3 and caspase-8 and then trigger mitochondria-mediated and caspase-dependent pathways in renal carcinoma cells [87]. YC-1 has been shown to downregulate several invasion-related signaling proteins, such as β-catenin, caveolin, Src, and epidermal growth factor receptor (EGFR), as well as multiple growth-related proteins, including 5'-AMP-activated protein kinase α (AMPKα), phospho-acetyl-CoA carboxylase (p-ACC), human epidermal growth factor receptor 2 (HER-2), and mammalian target of rapamycin (mTOR) in nasopharyngeal carcinoma [88]. Other anti-invasion mechanisms of YC-1 have also been identified in nasopharyngeal carcinoma (NPC) cells by reverse phase protein array [88]. Activation of beta-catenin signaling has also been evidenced to involve in inhibiting the proliferation and metastasis of hepatocellular carcinoma using combination therapy with local radiofrequency ablation and YC-1 [89]. Moreover, the previous study indicated that YC-1 has a potential effect to improve drug resistance by inhibiting multidrug-resistant protein resulting in decrease of P-glycoprotein (Pgp) efflux, whose effect is modulated by the NO-cGMP-PKG-ERK signaling pathway [90]. These observations revealed together that YC-1 exerts inhibitory effects in key signaling pathways essential for maintaining cancer or endothelial cell viability and may be developed as an antitumor agent on a broad spectrum of cancer types by facilitating apoptosis and suppressing tumor angiogenesis.

5.3 Neuroprotective and anti-inflammatory effects

The use of NO donors (e.g., NONOate) results in excessive NO production which may cause NO-induced axonal damage by inhibiting mitochondrial respiration, independent of cGMP [91]. YC-1 has been shown to protect white matter axons from NO toxicity. This axonoprotective action of YC-1 was unrelated to its activity on sGC but through a novel action on voltage-dependent Na^+ channels in the rat isolated optic nerve [92]. Lu et al. showed YC-1 inhibition of lipopolysaccharide (LPS)-induced iNOS and cyclooxygenase-2 (COX-2) expression as well as NF-κB activation, implying that YC-1 can be developed as an anti-inflammatory neuroprotective agent [93]. Chien et al. reported that YC-1 promoted learning behavior in Morris water maze and avoidance tests and YC-1 pretreatment reduced scopolamine-induced learning deficit. Thereby, the NO/cGMP/PKG pathway may be involved in the learning enhancement-based experiments with intracerebroventricular injection of L-NAME and PKG inhibitors [94]. Similarly, YC-1 can also improve age-related learning and memory dysfunction [95]. Furthermore, YC-1 may inhibit HIF-1α accumulation and VEGF production to protect blood-brain barrier against ischemia-/reperfusion-induced injury [96]. In addition, beneficial effect of YC-1 in ameliorating combined allergic rhinitis and asthma syndrome (CARAS) was demonstrated through reducing expressions of HIF-1α, NF-kB, and peroxisome proliferator-activated receptor α (PPARα) [97].

5.4 Optical protections

Therapeutic application of YC-1 on sepsis has been mentioned. After administration with YC-1, several LPS-stimulated modulations, such as NF-κB activation, iNOS expression, NO overproduction, and cytokine release, were markedly inhibited, thus improving survival rate of endotoxemic mice [98]. YC-1 has also been shown to inhibit HIF-1α-induced iNOS and VEGF expressions in various tissue models. Studies showed that YC-1 inhibited optical neovascularization in the pathological stages [99, 100]. Song noted that YC-1 could prevent laser-induced choroidal neovascularization by suppressing photocoagulation-mediated HIF-1 expression [99]. The pathological retinal neovascularization could also be inhibited by YC-1 through decreasing ischemia-induced expression of HIF-1 and its downstream angiogenic mediators (e.g., VEGF) in the ischemic retina. The physiological revascularization of the retinal vascular plexuses was enhanced by YC-1 *via* inhibiting iNOS expression at mRNA and protein levels [100]. Besides, it also has been reported that YC-1 alleviated macular edema in the animal model of laser-induced experimental central retinal vein occlusion by reducing several inflammatory or angiogenesis-related factors, such as interleukin-6 (IL-6), IL-8, VEGF, and HIF-1 [101].

5.5 Other activities

Wang and his colleagues evidenced that YC-1 inhibited bone resorption and induced extrinsic apoptosis of osteoclasts to reduce bone loss, which implied that YC-1 has potential application for use as an antiresorptive drug in postmenopausal osteoporosis [102]. Besides, YC-1 and its derivatives also have been mentioned to improve hepatic fibrosis, which mechanisms may be caused by inhibiting liver neutrophil infiltration as well as decreasing in TNF-α signaling and macrophage aggregation [103, 104].

6. Clinical significance of YC-1

Extensive studies have been conducted to explore possible systemic actions of YC-1 in disease models in animals to demonstrate that YC-1 has versatile physiological activities to be a potent candidate drug for a number of vascular disorders. In the cardiovascular and hematological systems, it has been reported that local extravascular administration of YC-1 could prevent neointima formation in a rat carotid artery model of balloon angioplasty [63, 64, 67]. In a study of experimental thrombosis model, YC-1 conferred beneficial effect through its anti-aggregating and pro-fibrinolytic effects [105]. BAY 63-2521 (riociguat™), a NO-independent but heme-dependent sGC stimulators like TC-1, is currently in clinical development for the treatment of pulmonary arterial hypertension with the only reported significant side effect to be a decrease in systemic arterial diastolic pressure [106, 107]. Similarly, intravenous administration of YC-1 has been shown to lower mean arterial blood pressure in normotensive and hypertensive rat [108]. For anticancer therapy, Lau et al. demonstrated that intraperitoneal injection of YC-1 enhances cisplatin chemosensitivity of hepatocellular carcinoma cells in nude mice xenograft tumor model, suggesting that YC-1 may be as an adjuvant agent for anticancer therapy [80]. Furthermore, oral administration of YC-1 can also decrease tumor mass in human renal cancer xenograft mice model [87]. In Morris water maze and avoidance test of mice, Chien et al. showed that YC-1 may be a good candidate for the improvement of learning and memory [94, 109]. Hwang et al. demonstrated that YC-1 can potentiate the relaxant responses of exogenous or endogenous NO through the elevation of cGMP in guinea-pig trachea [110]. The above in vivo studies all demonstrated the relevance of YC-1 in association with NO.

7. Conclusions

Accumulating evidences have shown that the administration of YC-1 may have beneficial pharmacological or physiological functions in diseased states for clinical applications. In the future, less toxic and more effective candidates would be the focus of further investigations through structural modification of YC-1 or its derivatives and better understanding of the molecular mechanisms of its actions.

Conflict of interest

The authors state that they have no conflict of interest.

Author details

Chieh-Hsi Wu[1*], Chun-Hsu Pan[1] and Ming-Jyh Sheu[2]

1 School of Pharmacy, Taipei Medical University, Taipei, Taiwan

2 School of Pharmacy, China Medical University, Taichung, Taiwan

*Address all correspondence to: chhswu@tmu.edu.tw

References

[1] Martin E, Davis K, Bian K, Lee YC, Murad F. Cellular signaling with nitric oxide and cyclic guanosine monophosphate. Seminars in Perinatology. 2000;**24**:2-6. DOI: 10.1016/S0146-0005(00)80045-2

[2] Murad F. Nitric oxide signaling: Would you believe that a simple free radical could be a second messenger, autacoid, paracrine substance, neurotransmitter, and hormone? Recent Progress in Hormone Research. 1998;**53**:43-59. discussion 59-60

[3] Deguchi T, Yoshioka M. L-Arginine identified as an endogenous activator for soluble guanylate cyclase from neuroblastoma cells. Journal of Biological Chemistry. 1982;**257**:10147-10151

[4] Hibbs JB Jr, Taintor RR, Vavrin Z. Macrophage cytotoxicity: Role for L-arginine deiminase and imino nitrogen oxidation to nitrite. Science. 1987;**235**:473-476. DOI: 10.1126/science.2432665

[5] Ignarro LJ, Buga GM, Wei LH, Bauer PM, Wu G, del Soldato P. Role of the arginine-nitric oxide pathway in the regulation of vascular smooth muscle cell proliferation. Proceedings of the National Academy of Sciences of the United States of America. 2001;**98**:4202-4208. DOI: 10.1073/pnas.071054698

[6] Bredt DS, Hwang PM, Glatt CE, Lowenstein C, Reed RR, Snyder SH. Cloned and expressed nitric oxide synthase structurally resembles cytochrome P-450 reductase. Nature. 1991;**351**:714-718. DOI: 10.1038/351714a0

[7] Nathan C, Xie QW. Nitric oxide synthases: Roles, tolls, and controls. Cell. 1994;**78**:915-918. DOI: 0092-8674(94)90266-6

[8] Janssens SP, Simouchi A, Quertermous T, Bloch DB, Bloch KD. Cloning and expression of a cDNA encoding human endothelium-derived relating factor/nitric oxide synthase. Journal of Biological Chemistry. 1992;**267**:22694

[9] Vallance P, Leone A, Calver A, Collier J, Moncada S. Accumulation of an endogenous inhibitor of nitric oxide synthesis in chronic renal failure. Lancet. 1992;**339**:572-575. DOI: 0140-6736(92)90865-Z

[10] Cooke JP. Does ADMA cause endothelial dysfunction? Arteriosclerosis, Thrombosis, and Vascular Biology. 2000;**20**:2032-2037. DOI: 10.1161/01.ATV.20.9.2032

[11] Ueda S, Yamagishi S, Matsumoto Y, Fukami K, Okuda S. Asymmetric dimethylarginine (ADMA) is a novel emerging risk factor for cardiovascular disease and the development of renal injury in chronic kidney disease. Clinical and Experimental Nephrology. 2007;**11**:115-121. DOI: 10.1007/s10157-007-0471-x

[12] Murad F. Discovery of some of the biological effects of nitric oxide and its role in cell signaling. Bioscience Reports. 1999;**19**:133-154. DOI: 10.1023/A:1020265417394

[13] Bian K, Murad F. Nitric oxide (NO)—Biogeneration, regulation, and relevance to human diseases. Frontiers in Bioscience. 2003;**8**:d264-d278. DOI: 10.2741/997

[14] Murad F. Shattuck Lecture. Nitric oxide and cyclic GMP in cell signaling and drug development. New England Journal of Medicine. 2006;**355**:2003-2011. DOI: 10.1056/NEJMsa063904

[15] Cai H, Harrison DG. Endothelial dysfunction in cardiovascular diseases:

The role of oxidant stress. Circulation Research. 2000;**87**:840-844. DOI: 10.1161/res.87.10.840

[16] Wilcox CS. Redox regulation of the afferent arteriole and tubuloglomerular feedback. Acta Physiologica Scandinavica. 2003;**179**:217-223. DOI: 10.1046/j.0001-6772.2003.01205.x

[17] Bian K, Doursout MF, Murad F. Vascular system: Role of nitric oxide in cardiovascular diseases. Journal of Clinical Hypertension. 2008;**10**:304-310. DOI: 10.1111/j.1751-7176.2008.06632.x

[18] Ullrich V, Kissner R. Redox signaling: Bioinorganic chemistry at its best. Journal of Inorganic Biochemistry. 2006;**100**:2079-2086. DOI: 10.1016/j.jinorgbio.2006.09.019

[19] Calver A, Collier J, Leone A, Moncada S, Vallance P. Effect of local intra-arterial asymmetric dimethylarginine (ADMA) on the forearm arteriolar bed of healthy volunteers. Journal of Human Hypertension. 1993;7:193-194

[20] Takiuchi S, Fujii H, Kamide K, Horio T, Nakatani S, Hiuge A, et al. Plasma asymmetric dimethylarginine and coronary and peripheral endothelial dysfunction in hypertensive patients. American Journal of Hypertension. 2004;**17**:802-808. DOI: 10.1016/j.amjhyper.2004.05.005

[21] Perticone F, Sciacqua A, Maio R, Perticone M, Maas R, Boger RH, et al. Asymmetric dimethylarginine, L-arginine, and endothelial dysfunction in essential hypertension. Journal of the American College of Cardiology. 2005;**46**:518-523. DOI: 10.1016/j.jacc.2005.04.040

[22] Perticone F, Ceravolo R, Pujia A, Ventura G, Iacopino S, Scozzafava A, et al. Prognostic significance of endothelial dysfunction in hypertensive patients. Circulation. 2001;**104**:191-196. DOI: 10.1161/01.CIR.104.2.191

[23] Böger RH, Lentz SR, Bode-Böger SM, Knapp HR, Haynes WG. Elevation of asymmetrical dimethylarginine may mediate endothelial dysfunction during experimental hyperhomocyst(e)inaemia in humans. Clinical Science. 2001;**100**:161-167. DOI: 10.1042/cs1000161

[24] Meigs JB, Jacques PF, Selhub J, Singer DE, Nathan DM, Rifai N, et al. Fasting plasma homocysteine levels in the insulin resistance syndrome: The Framingham offspring study. Diabetes Care. 2001;**24**:1403-1410. DOI: 10.2337/diacare.24.8.1403

[25] Stuhlinger MC, Tsao PS, Her JH, Kimoto M, Balint RF, Cooke JP. Homocysteine impairs the nitric oxide synthase pathway: Role of asymmetric dimethylarginine. Circulation. 2001;**104**:2569-2575. DOI: 10.1161/hc4601.098514

[26] Tyagi N, Sedoris KC, Steed M, Ovechkin AV, Moshal KS, Tyagi SC. Mechanisms of homocysteine-induced oxidative stress. American Journal of Physiology. Heart and Circulatory Physiology. 2005;**289**:H2649-H2656. DOI: 10.1152/ajpheart.00548.2005

[27] Frey D, Braun O, Briand C, Vasak M, Grutter MG. Structure of the mammalian NOS regulator dimethylarginine dimethylaminohydrolase: A basis for the design of specific inhibitors. Structure. 2006;**14**:901-911. DOI: 10.1016/j.str.2006.03.006

[28] Leiper JM, Santa Maria J, Chubb A, MacAllister RJ, Charles IG, Whitley GS, et al. Identification of two human dimethylarginine dimethylaminohydrolases with distinct tissue distributions and homology with microbial arginine deiminases.

Biochemical Journal. 1999;**343** (Pt 1):209-214. DOI: 10.1042/bj3430209

[29] Nijveldt RJ, Teerlink T, van Guldener C, Prins HA, van Lambalgen AA, Stehouwer CD, et al. Handling of asymmetrical dimethylarginine and symmetrical dimethylarginine by the rat kidney under basal conditions and during endotoxaemia. Nephrology, Dialysis, Transplantation. 2003;**18**:2542-2550. DOI: 10.1093/ndt/gfg452

[30] Ito A, Tsao PS, Adimoolam S, Kimoto M, Ogawa T, Cooke JP. Novel mechanism for endothelial dysfunction: Dysregulation of dimethylarginine dimethylaminohydrolase. Circulation. 1999;**99**:3092-3095. DOI: 10.1161/01.CIR.99.24.3092

[31] Böger RH, Sydow K, Borlak J, Thum T, Lenzen H, Schubert B, et al. LDL cholesterol upregulates synthesis of asymmetrical dimethylarginine in human endothelial cells: Involvement of S-adenosylmethionine-dependent methyltransferases. Circulation Research. 2000;**87**:99-105. DOI: 10.1161/res.87.2.99

[32] Harada-Shiba M, Kinoshita M, Kamido H, Shimokado K. Oxidized low density lipoprotein induces apoptosis in cultured human umbilical vein endothelial cells by common and unique mechanisms. Journal of Biological Chemistry. 1998;**273**:9681-9687. DOI: 10.1074/jbc.273.16.9681

[33] Ma FX, Zhou B, Chen Z, Ren Q, Lu SH, Sawamura T, et al. Oxidized low density lipoprotein impairs endothelial progenitor cells by regulation of endothelial nitric oxide synthase. Journal of Lipid Research. 2006;**47**:1227-1237. DOI: 10.1194/jlr.M500507-JLR200

[34] Chin JH, Azhar S, Hoffman BB. Inactivation of endothelial derived relaxing factor by oxidized lipoproteins. Journal of Clinical Investigation. 1992;**89**:10-18. DOI: 10.1172/JCI115549

[35] Okyay K, Cengel A, Sahinarslan A, Tavil Y, Turkoglu S, Biberoglu G, et al. Plasma asymmetric dimethylarginine and L-arginine levels in patients with cardiac syndrome X. Coronary Artery Disease. 2007;**18**:539-544. DOI: 10.1097/MCA.0b013e3282f08ece

[36] Sen N, Poyraz F, Tavil Y, Yazici HU, Turfan M, Hizal F, et al. Carotid intima-media thickness in patients with cardiac syndrome X and its association with high circulating levels of asymmetric dimethylarginine. Atherosclerosis. 2009;**204**:e82-e85. DOI: 10.1016/j.atherosclerosis.2008.09.029

[37] Zoccali C, Bode-Boger S, Mallamaci F, Benedetto F, Tripepi G, Malatino L, et al. Plasma concentration of asymmetrical dimethylarginine and mortality in patients with end-stage renal disease: A prospective study. Lancet. 2001;**358**:2113-2117. DOI: 10.1016/S0140-6736(01)07217-8

[38] Kielstein JT, Boger RH, Bode-Boger SM, Frolich JC, Haller H, Ritz E, et al. Marked increase of asymmetric dimethylarginine in patients with incipient primary chronic renal disease. Journal of the American Society of Nephrology. 2002;**13**:170-176

[39] Fliser D, Kronenberg F, Kielstein JT, Morath C, Bode-Boger SM, Haller H, et al. Asymmetric dimethylarginine and progression of chronic kidney disease: The mild to moderate kidney disease study. Journal of the American Society of Nephrology. 2005;**16**:2456-2461. DOI: 10.1681/ASN.2005020179

[40] Schmidt R, Fazekas F, Hayn M, Schmidt H, Kapeller P, Roob G, et al. Risk factors for microangiopathy-related cerebral damage in the Austrian stroke prevention study. Journal of the Neurological Sciences. 1997;**152**:15-21. DOI: 10.1016/S0022-510X(97)00137-8

[41] Englund E. Neuropathology of white matter lesions in vascular cognitive impairment. Cerebrovascular Diseases. 2002;**13**(Suppl 2):11-15. DOI: 10.1159/000049144

[42] Shibata H, Nabika T, Moriyama H, Masuda J, Kobayashi S. Correlation of NO metabolites and 8-iso-prostaglandin F2α with periventricular hyperintensity severity. Arteriosclerosis, Thrombosis, and Vascular Biology. 2004;**24**:1659-1663. DOI: 10.1161/01.ATV.0000137415.67349.3c

[43] Wright CB, Paik MC, Brown TR, Stabler SP, Allen RH, Sacco RL, et al. Total homocysteine is associated with white matter hyperintensity volume: The Northern Manhattan Study. Stroke. 2005;**36**:1207-1211. DOI: 10.1161/01.STR.0000165923.02318.22

[44] Hassan A, Gormley K, O'Sullivan M, Knight J, Sham P, Vallance P, et al. Endothelial nitric oxide gene haplotypes and risk of cerebral small-vessel disease. Stroke. 2004;**35**:654-659. DOI: 10.1161/01.STR.0000117238.75736.53

[45] Nasreen S, Nabika T, Shibata H, Moriyama H, Yamashita K, Masuda J, et al. T-786C polymorphism in endothelial NO synthase gene affects cerebral circulation in smokers: Possible gene-environmental interaction. Arteriosclerosis, Thrombosis, and Vascular Biology. 2002;**22**:605-610. DOI: 10.1161/01.ATV.20.9.2032

[46] Neufeld AH. Pharmacologic neuroprotection with an inhibitor of nitric oxide synthase for the treatment of glaucoma. Brain Research Bulletin. 2004;**62**:455-459. DOI: 10.1016/j.brainresbull.2003.07.005

[47] Kuo SC, Lee FY, Teng CM. 1-(Substituted benzyl)-3-(substituted aryl)-condensed pyrazole derivatives and processes of making the same. US Patent 5574168

[48] Lee FY, Lien JC, Huang LJ, Huang TM, Tsai SC, Teng CM, et al. Synthesis of 1-benzyl-3-(5′-hydroxymethyl-2′-furyl)indazole analogues as novel antiplatelet agents. Journal of Medicinal Chemistry. 2001;**44**:3746-3749. DOI: 10.1021/jm010001h

[49] Ko FN, Wu CC, Kuo SC, Lee FY, Teng CM. YC-1, a novel activator of platelet guanylate cyclase. Blood 1994;**84**:4226-4233

[50] Wu CC, Ko FN, Kuo SC, Lee FY, Teng CM. YC-1 inhibited human platelet aggregation through NO-independent activation of soluble guanylate cyclase. British Journal of Pharmacology. 1995;**116**:1973-1978. DOI: 10.1111/j.1476-5381.1995.tb16400.x

[51] Stasch JP, Hobbs AJ. NO-independent, haem-dependent soluble guanylate cyclase stimulators. Handbook of Experimental Pharmacology. 2009:277-308. DOI: 10.1007/978-3-540-68964-5_13

[52] Friebe A, Schultz G, Koesling D. Sensitizing soluble guanylyl cyclase to become a highly CO-sensitive enzyme. EMBO Journal. 1996;**15**:6863-6868

[53] Martin E, Lee YC, Murad F. YC-1 activation of human soluble guanylyl cyclase has both heme-dependent and heme-independent components. Proceedings of the National Academy of Sciences of the United States of America. 2001;**98**:12938-12942. DOI: 10.1073/pnas.231486198

[54] Schmidt K, Schrammel A, Koesling D, Mayer B. Molecular mechanisms involved in the synergistic activation of soluble guanylyl cyclase by YC-1 and nitric oxide in endothelial cells. Molecular Pharmacology. 2001;**59**:220-224. DOI: 10.1124/mol.59.2.220

[55] Stasch JP, Becker EM, Alonso-Alija C, Apeler H, Dembowsky K, Feurer A, et al.

NO-independent regulatory site on soluble guanylate cyclase. Nature. 2001;**410**:212-215. DOI: 10.1038/35065611

[56] Mulsch A, Bauersachs J, Schafer A, Stasch JP, Kast R, Busse R. Effect of YC-1, an NO-independent, superoxide-sensitive stimulator of soluble guanylyl cyclase, on smooth muscle responsiveness to nitrovasodilators. British Journal of Pharmacology. 1997;**120**:681-689. DOI: 10.1038/sj.bjp.0700982

[57] Wohlfart P, Malinski T, Ruetten H, Schindler U, Linz W, Schoenafinger K, et al. Release of nitric oxide from endothelial cells stimulated by YC-1, an activator of soluble guanylyl cyclase. British Journal of Pharmacology. 1999;**128**:1316-1322. DOI: 10.1038/sj.bjp.0702921

[58] Friebe A, Mullershausen F, Smolenski A, Walter U, Schultz G, Koesling D. YC-1 potentiates nitric oxide- and carbon monoxide-induced cyclic GMP effects in human platelets. Molecular Pharmacology. 1998;**54**:962-967. DOI: 10.1124/mol.54.6.962

[59] Galle J, Zabel U, Hubner U, Hatzelmann A, Wagner B, Wanner C, et al. Effects of the soluble guanylyl cyclase activator, YC-1, on vascular tone, cyclic GMP levels and phosphodiesterase activity. British Journal of Pharmacology. 1999;**127**:195-203. DOI: 10.1038/sj.bjp.0702495

[60] Aszódi A, Pfeifer A, Ahmad M, Glauner M, Zhou XH, Ny L, et al. The vasodilator-stimulated phosphoprotein (VASP) is involved in cGMP- and cAMP-mediated inhibition of agonist-induced platelet aggregation, but is dispensable for smooth muscle function. EMBO Journal. 1999;**18**:37-48. DOI: 10.1093/emboj/18.1.37

[61] Becker EM, Schmidt P, Schramm M, Schroder H, Walter U, Hoenicka M, et al. The vasodilator-stimulated phosphoprotein (VASP): Target of YC-1 and nitric oxide effects in human and rat platelets. Journal of Cardiovascular Pharmacology. 2000;**35**:390-397. DOI: 10.1097/00005344-200003000-00007

[62] Yu SM, Cheng ZJ, Guh JH, Lee FY, Kuo SC. Mechanism of anti-proliferation caused by YC-1, an indazole derivative, in cultured rat A10 vascular smooth-muscle cells. Biochemical Journal. 1995;**306** (Pt 3):787-792. DOI: 10.1042/bj3060787

[63] Tulis DA, Durante W, Peyton KJ, Chapman GB, Evans AJ, Schafer AI. YC-1, a benzyl indazole derivative, stimulates vascular cGMP and inhibits neointima formation. Biochemical and Biophysical Research Communications. 2000;**279**:646-652. DOI: 10.1006/bbrc.2000.3942

[64] Wu CH, Chang WC, Chang GY, Kuo SC, Teng CM. The inhibitory mechanism of YC-1, a benzyl indazole, on smooth muscle cell proliferation: An in vitro and in vivo study. Journal of Pharmacological Sciences. 2004;**94**:252-260. DOI: 10.1254/jphs.94.252

[65] Lincoln TM, Dey NB, Boerth NJ, Cornwell TL, Soff GA. Nitric oxide-cyclic GMP pathway regulates vascular smooth muscle cell phenotypic modulation: Implications in vascular diseases. Acta Physiologica Scandinavica. 1998;**164**:507-515. DOI: 10.1111/j.1365-201X.1998.tb10700.x

[66] Kaul S, Cercek B, Rengstrom J, Xu XP, Molloy MD, Dimayuga P, et al. Polymeric-based perivascular delivery of a nitric oxide donor inhibits intimal thickening after balloon denudation arterial injury: Role of nuclear factor-kappaB. Journal of the American College of Cardiology. 2000;**35**:493-501. DOI: 10.1016/S0735-1097(99)00543-4

[67] Liu YN, Pan SL, Peng CY, Guh JH, Huang DM, Chang YL, et al. YC-1

[3-(5′-hydroxymethyl-2′-furyl)-1-benzyl indazole] inhibits neointima formation in balloon-injured rat carotid through suppression of expressions and activities of matrix metalloproteinases 2 and 9. Journal of Pharmacology and Experimental Therapeutics. 2006;**316**:35-41. DOI: 10.1124/jpet.105.090563

[68] Liu YN, Pan SL, Peng CY, Huang DY, Guh JH, Kuo SC, et al. YC-1 induces heat shock protein 70 expression and prevents oxidized LDL-mediated apoptosis in vascular smooth muscle cells. Shock. 2008;**30**:274-279. DOI: 10.1097/shk.0b013e318162c63a

[69] Lam KK, Cheng PY, Lee YM, Liu YP, Ding C, Liu WH, et al. The role of heat shock protein 70 in the protective effect of YC-1 on heat stroke rats. European Journal of Pharmacology. 2013;**699**: 67-73. DOI: 10.1016/j.ejphar.2012.11.044

[70] Keswani AN, Peyton KJ, Durante W, Schafer AI, Tulis DA. The cyclic GMP modulators YC-1 and zaprinast reduce vessel remodeling through antiproliferative and proapoptotic effects. Journal of Cardiovascular Pharmacology and Therapeutics. 2009;**14**:116-124. DOI: 10.1177/1074248409333266

[71] Hsu HK, Juan SH, Ho PY, Liang YC, Lin CH, Teng CM, et al. YC-1 inhibits proliferation of human vascular endothelial cells through a cyclic GMP-independent pathway. Biochemical Pharmacology. 2003;**66**:263-271. DOI: 10.1016/S0006-2952(03)00244-2

[72] Huh JW, Kim SY, Lee JH, Lee YS. YC-1 attenuates hypoxia-induced pulmonary arterial hypertension in mice. Pulmonary Pharmacology & Therapeutics. 2011;**24**:638-646. DOI: 10.1016/j.pupt.2011.09.003

[73] Chun YS, Yeo EJ, Choi E, Teng CM, Bae JM, Kim MS, et al. Inhibitory effect of YC-1 on the hypoxic induction of erythropoietin and vascular endothelial growth factor in Hep3B cells. Biochemical Pharmacology. 2001;**61**:947-954. DOI: 10.1016/S0006-2952(01)00564-0

[74] Shin DH, Kim JH, Jung YJ, Kim KE, Jeong JM, Chun YS, et al. Preclinical evaluation of YC-1, a HIF inhibitor, for the prevention of tumor spreading. Cancer Letters. 2007;**255**:107-116. DOI: 10.1016/j.canlet.2007.03.026

[75] Sun HL, Liu YN, Huang YT, Pan SL, Huang DY, Guh JH, et al. YC-1 inhibits HIF-1 expression in prostate cancer cells: Contribution of Akt/NF-kappaB signaling to HIF-1alpha accumulation during hypoxia. Oncogene. 2007;**26**:3941-3951. DOI: 10.1038/sj.onc.1210169

[76] Zhao Q , Du J, Gu H, Teng X, Zhang Q , Qin H, et al. Effects of YC-1 on hypoxia-inducible factor 1-driven transcription activity, cell proliferative vitality, and apoptosis in hypoxic human pancreatic cancer cells. Pancreas. 2007;**34**:242-247. DOI: 10.1097/01.mpa.0000250135.95144.b6

[77] Moon SY, Chang HW, Roh JL, Kim GC, Choi SH, Lee SW, et al. Using YC-1 to overcome the radioresistance of hypoxic cancer cells. Oral Oncology. 2009;**45**:915-919. DOI: 10.1016/j.oraloncology.2009.04.005

[78] Lau CK, Yang ZF, Lam CT, Tam KH, Poon RT, Fan ST. Suppression of hypoxia inducible factor-1alpha (HIF-1alpha) by YC-1 is dependent on murine double minute 2 (Mdm2). Biochemical and Biophysical Research Communications. 2006;**348**:1443-1448. DOI: 10.1016/j.bbrc.2006.08.015

[79] Wakiyama K, Kitajima Y, Tanaka T, Kaneki M, Yanagihara K, Aishima S, et al. Low-dose YC-1 combined with glucose and insulin selectively induces apoptosis in hypoxic gastric carcinoma cells by inhibiting anaerobic glycolysis.

Scientific Reports. 2017;7:12653. DOI: 10.1038/s41598-017-12929-9

[80] Lau CK, Yang ZF, Lam SP, Lam CT, Ngai P, Tam KH, et al. Inhibition of Stat3 activity by YC-1 enhances chemo-sensitivity in hepatocellular carcinoma. Cancer Biology & Therapy. 2007;**6**:1900-1907. DOI: 10.4161/cbt.6.12.4970

[81] Lee SJ, Kim YJ, Lee CS, Bae J. Combined application of camptothecin and the guanylate cyclase activator YC-1: Impact on cell death and apoptosis-related proteins in ovarian carcinoma cell lines. Chemico-Biological Interactions. 2009;**181**:185-192. DOI: 10.1016/j.cbi.2009.05.013

[82] Huang YT, Pan SL, Guh JH, Chang YL, Lee FY, Kuo SC, et al. YC-1 suppresses constitutive nuclear factor-kappaB activation and induces apoptosis in human prostate cancer cells. Molecular Cancer Therapeutics. 2005;**4**:1628-1635. DOI: 10.1158/1535-7163.MCT-05-0090

[83] Pan SL, Guh JH, Peng CY, Wang SW, Chang YL, Cheng FC, et al. YC-1 [3-(5′-hydroxymethyl-2′-furyl)-1-benzyl indazole] inhibits endothelial cell functions induced by angiogenic factors in vitro and angiogenesis in vivo models. Journal of Pharmacology and Experimental Therapeutics. 2005;**314**:35-42. DOI: 10.1124/jpet.105.085126

[84] Wang SW, Pan SL, Guh JH, Chen HL, Huang DM, Chang YL, et al. YC-1 [3-(5'-hydroxymethyl-2′-furyl)-1-benzyl indazole] exhibits a novel antiproliferative effect and arrests the cell cycle in G0-G1 in human hepatocellular carcinoma cells. Journal of Pharmacology and Experimental Therapeutics. 2005;**312**:917-925. DOI: 10.1124/jpet.104.077230

[85] Yeo EJ, Ryu JH, Chun YS, Cho YS, Jang IJ, Cho H, et al. YC-1 induces S cell cycle arrest and apoptosis by activating checkpoint kinases. Cancer Research. 2006;**66**:6345-6352. DOI: 10.1158/0008-5472.CAN-05-4460

[86] Lee MR, Lin C, Lu CC, Kuo SC, Tsao JW, Juan YN, et al. YC-1 induces G0/G1 phase arrest and mitochondria-dependent apoptosis in cisplatin-resistant human oral cancer CAR cells. Biomedicine (Taipei). 2017;7:12. DOI: 10.1051/bmdcn/2017070205

[87] Wu SY, Pan SL, Chen TH, Liao CH, Huang DY, Guh JH, et al. YC-1 induces apoptosis of human renal carcinoma A498 cells in vitro and in vivo through activation of the JNK pathway. British Journal of Pharmacology. 2008;**155**:505-513. DOI: 10.1038/bjp.2008.292

[88] Hong B, Lui VW, Hui EP, Lu Y, Leung HS, Wong EY, et al. Reverse phase protein array identifies novel anti-invasion mechanisms of YC-1. Biochemical Pharmacology. 2010;**79**:842-852. DOI: 10.1016/j.bcp.2009.10.021

[89] Zhao W, Li X, Li Z. Combination therapy with local radiofrequency ablation and YC-1 inhibits the proliferation and metastasis of hepatocellular carcinoma through activating beta-catenin signaling. Die Pharmazie. 2016;**71**:524-529. DOI: 10.1691/ph.2016.6602

[90] Hung CC, Liou HH. YC-1, a novel potential anticancer agent, inhibit multidrug-resistant protein via cGMP-dependent pathway. Investigational New Drugs. 2011;**29**:1337-1346. DOI: 10.1007/s10637-010-9496-1

[91] Garthwaite G, Goodwin DA, Batchelor AM, Leeming K, Garthwaite J. Nitric oxide toxicity in CNS white matter: An in vitro study using rat optic nerve. Neuroscience. 2002;**109**:145-155. DOI: 10.1016/S0306-4522(01)00447-X

[92] Garthwaite G, Goodwin DA, Neale S, Riddall D, Garthwaite J. Soluble guanylyl

cyclase activator YC-1 protects white matter axons from nitric oxide toxicity and metabolic stress, probably through Na(+) channel inhibition. Molecular Pharmacology. 2002;**61**:97-104. DOI: 10.1124/mol.61.1.97

[93] Lu DY, Tang CH, Liou HC, Teng CM, Jeng KC, Kuo SC, et al. YC-1 attenuates LPS-induced proinflammatory responses and activation of nuclear factor-kappaB in microglia. British Journal of Pharmacology. 2007;**151**:396-405. DOI: 10.1038/sj.bjp.0707187

[94] Chien WL, Liang KC, Teng CM, Kuo SC, Lee FY, Fu WM. Enhancement of learning behaviour by a potent nitric oxide-guanylate cyclase activator YC-1. European Journal of Neuroscience. 2005;**21**:1679-1688. DOI: 10.1111/j.1460-9568.2005.03993.x

[95] Komsuoglu Celikyurt I, Utkan T, Ozer C, Gacar N, Aricioglu F. Effects of YC-1 on learning and memory functions of aged rats. Medical Science Monitor Basic Research. 2014;**20**:130-137. DOI: 10.12659/MSMBR.891064

[96] Yeh WL, Lu DY, Lin CJ, Liou HC, Fu WM. Inhibition of hypoxia-induced increase of blood-brain barrier permeability by YC-1 through the antagonism of HIF-1alpha accumulation and VEGF expression. Molecular Pharmacology. 2007;**72**:440-449. DOI: 10.1124/mol.107.036418

[97] Wang X, Liu C, Wu L, Zhu S. Potent ameliorating effect of hypoxia-inducible factor 1alpha (HIF-1alpha) antagonist YC-1 on combined allergic rhinitis and asthma syndrome (CARAS) in rats. European Journal of Pharmacology. 2016;**788**:343-350. DOI: 10.1016/j.ejphar.2016.07.040

[98] Pan SL, Guh JH, Peng CY, Chang YL, Cheng FC, Chang JH, et al. A potential role of YC-1 on the inhibition of cytokine release in peripheral blood mononuclear leukocytes and endotoxemic mouse models. Thrombosis and Haemostasis. 2005;**93**:940-948. DOI: 10.1267/THRO05050940

[99] Song SJ, Chung H, Yu HG. Inhibitory effect of YC-1, 3-(5′-hydroxymethyl-2′-furyl)-1-benzylindazole, on experimental choroidal neovascularization in rat. Ophthalmic Research. 2008;**40**:35-40. DOI: 10.1159/000111157

[100] Deniro M, Al-Halafi A, Al-Mohanna FH, Alsmadi O, Al-Mohanna FA. Pleiotropic effects of YC-1 selectively inhibits pathological retinal neovascularization and promotes physiological revascularization in a mouse model of oxygen-induced retinopathy. Molecular Pharmacology. 2009;**77**:348-367. DOI: 10.1124/mol.109.061366

[101] Yan Z, An J, Shang Q , Zhou N, Ma J. YC-1 inhibits VEGF and inflammatory mediators expression on experimental central retinal vein occlusion in rhesus monkey. Current Eye Research. 2018;**43**:526-533. DOI: 10.1080/02713683.2018.1426102

[102] Wang JW, Yeh CB, Chou SJ, Lu KC, Chu TH, Chen WY, et al. YC-1 alleviates bone loss in ovariectomized rats by inhibiting bone resorption and inducing extrinsic apoptosis in osteoclasts. Journal of Bone and Mineral Metabolism. 2018;**36**:508-518. DOI: 10.1007/s00774-017-0866-z

[103] Lee TY, Leu YL, Wen CK. Modulation of HIF-1alpha and STAT3 signaling contributes to anti-angiogenic effect of YC-1 in mice with liver fibrosis. Oncotarget. 2017;**8**:86206-86216. DOI: 10.18632/oncotarget.21039

[104] Xiao J, Jin C, Liu Z, Guo S, Zhang X, Zhou X, et al. The design, synthesis, and biological evaluation of novel

YC-1 derivatives as potent anti-hepatic fibrosis agents. Organic & Biomolecular Chemistry. 2015;**13**:7257-7264. DOI: 10.1039/c5ob00710k

[105] Teng CM, Wu CC, Ko FN, Lee FY, Kuo SC. YC-1, a nitric oxide-independent activator of soluble guanylate cyclase, inhibits platelet-rich thrombosis in mice. European Journal of Pharmacology. 1997;**320**:161-166. DOI: 10.1016/S0014-2999(96)00911-9

[106] Belik J. Riociguat, an oral soluble guanylate cyclase stimulator for the treatment of pulmonary hypertension. Current Opinion in Investigational Drugs. 2009;**10**:971-979

[107] Grimminger F, Weimann G, Frey R, Voswinckel R, Thamm M, Bolkow D, et al. First acute haemodynamic study of soluble guanylate cyclase stimulator riociguat in pulmonary hypertension. European Respiratory Journal. 2009;**33**:785-792. DOI: 10.1183/09031936.00039808

[108] Rothermund L, Friebe A, Paul M, Koesling D, Kreutz R. Acute blood pressure effects of YC-1-induced activation of soluble guanylyl cyclase in normotensive and hypertensive rats. British Journal of Pharmacology. 2000;**130**:205-208. DOI. 10.1038/sj.bjp.0703320

[109] Chien WL, Liang KC, Fu WM. Enhancement of active shuttle avoidance response by the NO-cGMP-PKG activator YC-1. European Journal of Pharmacology. 2008;**590**:233-240. DOI: 10.1016/j.ejphar.2008.06.040

[110] Hwang TL, Wu CC, Teng CM. YC-1 potentiates nitric oxide-induced relaxation in guinea-pig trachea. British Journal of Pharmacology. 1999;**128**:577-584. DOI: 10.1038/sj.bjp.0702830

Section 2

Vascular Diseases and Clinical Applications

Chapter 6

Giant Cell Arteritis: Current Advances in Pathogenesis and Treatment

Marília A. Dagostin and Rosa M.R. Pereira

Abstract

Giant cell arteritis (GCA) is the most common vasculitis in adults, with the incidence increasing with the advancing age. The aorta and its branches, especially the carotid extracranial branches, are the classic targets of inflammation in GCA. Visual loss, upper limb ischemia, and stroke are complications described. Suspicion of GCA is a medical emergency, and patients need to be quickly diagnosed/treated to prevent irreversible damage. Headache is the most common symptom, and a new-onset headache in older adults should always raise the suspicion of GCA. Patients may also present with scalp tenderness or tongue/jaw pain. GCA is often found to be the cause of an obscure-origin fever in older patients. A positive temporal artery biopsy is considered the gold standard for the diagnosis, but imaging techniques enable the assessment of cranial and extracranial arteries and the aorta. Ultrasound of temporal arteries is recommended and noncompressible "halo" sign is the typical finding. PET, MRI, or CT may be useful for the detection of the disease in the aorta and other vessels. The treatment must be started promptly with prednisone 1 mg/kg/day. When visual symptoms/unilateral visual loss is present, methylprednisolone pulse is recommended. Methotrexate, leflunomide and tocilizumab may be effective and well-tolerated glucocorticoid-sparing agents in GCA. Cardiovascular diseases are the leading causes of death in patients.

Keywords: giant cell arteritis, T17, IL-6, pathogenesis, treatment, tocilizumab

1. Introduction

The first description of giant cell arteritis dates from 1890 by Hutchinson, who described an 80-year-old man with painful and inflamed temporal arteries which prevented him from wearing his hat [1]. Forty-seven years later Horton, Magath, and Brown described similar cases and called the syndrome temporal arteritis [2]. Originally thought to be a localized, self-limiting, and benign disorder, inflammation of the temporal arteries is now recognized as part of a widespread arteritis which untreated can lead to blindness and death.

Giant cell arteritis (GCA), previously called temporal arteritis and also known as Horton's disease, is defined by the 2012 Chapel Hill Consensus Conference as "arteritis, often granulomatous, usually affecting the aorta and/or its major branches, with a predilection for the branches of the carotid and vertebral arteries; often involves the temporal artery; onset usually in patients older than 50; often associated with polymyalgia rheumatica" [3]. It is the most common primary systemic vasculitis in adults, mostly seen in North America and Western Europe, with the incidence increasing with the advancing age. Women are more affected than men in a 2.5:1 ratio. GCA classically targets large vessels with predominance for the aorta and its branches. Arterial inflammation may lead to vascular damage which can result in stenosis, occlusions, and even aneurysms. Therefore, this condition is related to serious loss of function including visual loss, upper limb ischemia, and stroke. Suspicion of giant cell arteritis is a medical emergency, and patients need to be quickly diagnosed and treated to prevent irreversible consequences of vessel inflammation.

2. Pathogenesis

The immune and pathogenetic pathways responsible for the inflammation on the arterial walls in GCA are not fully understood yet. As in other autoimmune diseases, it is believed to be an environmental-triggered response occurring in genetic-predisposed individuals. The fact that it only affects older patients suggests that age-related damage on the vessel walls also plays a role in the development of the arteritis [4].

There is evidence of a cyclic pattern and yearly incidence of the onset of GCA, which led to the search for an environmental agent responsible for the initialization of the immune response. Many bacterial agents and viruses have been under research (*Chlamydia pneumonia*, *Burkholderia*, *cytomegalovirus*, *erythrovirus B19*, *herpes simplex*, and *parainfluenzae 1*, among others), but the studies failed in finding a causal correlation between infections and GCA so far [4].

GCA is associated with the major histocompatibility class II (MHC-II), particularly with HLA-DRB1*04 alleles [5]. Outside the MHC region, variants on the PTPN22 locus and other genes related to vascular response to inflammation and vascular remodeling, such as plasminogen and prolyl 4-hydroxylase subunit alpha 2, also increased the risk of GCA.

Age-related damage on the arterial wall also has a role on the pathogenesis. There are biochemical and structural modifications in the vessel leading to loss of self-tolerance. Differences in the DNA methylation level of several genes have been reported in temporal arteries from GCA patients comparing with non-GCA controls.

There is evidence that the inflammation starts in the adventitia and progresses to the other layers of the arterial wall, culminating in transmural damage. Activated dendritic cells (DCs) have a central role in the immune response of GCA. These cells are present in the adventitia and express Toll-like receptors, which are activated via pathogen-associated molecular patterns (PAMPs) or microorganism-associated molecular patterns (MAMPs). The activation of DC breaks immune tolerance and renders the arteries, considered otherwise an immune privileged site, susceptible to inflammatory injury. DCs activate CD4+ T lymphocytes through co-stimulatory molecules (CD80 and CD86) and class II MHC. DC depletion in mice models with GCA strongly decreased vasculitis lesions, emphasizing its importance on the immune response.

Activated DCs produce cytokines such as IL-6, IL-18, IL-23, IL-32, and IL-33, which are chemotactic for T lymphocytes. T cells that infiltrate the temporal arteries from GCA patients are enriched Th1, Th17, and Th9 cells. Th1-response polarization via IL-12 synthesizes IFN-γ, and Th17 cells produce IL-17. While Th17 cells are inhibited by glucocorticoids, the Th1 response is not, being this last one implicated in sustaining chronic disease activity in GCA [6].

IFN-γ seems to be important for the development of vasculitis. The cytokine panel described above is found in both GCA patients and PMR patients without

GCA, but INF-γ is only present in individuals with GCA. IFN-γ expression, in fact, is associated with increased risk of ischemic complications. Vascular smooth muscle cells, induced by IFN-γ, produce chemokines (CCL2, CXCL9, CXCL10, and CXCL11), leading to the recruitment of monocytes that merge to form multinucleated giant cells, the hallmark of GCA. The chemokines recruit more immune cells, amplifying the inflammatory response. Monocytes differentiate into macrophages in the arterial wall and produce IL-6, IL-1β, and TNF-α, responsible for the systemic inflammatory response which is characteristic of GCA. Toxic substances for the arteries are also produced by macrophages, such as reactive oxygen species, matrix metalloproteinase-2 (MMP-2), and MMP-9, which destroy cellular matrix proteins and cause destruction of the media and digestion of the internal elastic lamina.

Th17 cells appear to be important in the initial stages of the disease, producing IL-17, IL-21, IL-22, and CCL20. IL-17 leads to the recruitment of macrophages, while IL-21 enhances the differentiation of cytotoxic cells, and IL-22 mediates hepatocyte stimulation and acute-phase amplification. CCL20 facilitates the recruitment of more DCs and T cells.

The vascular smooth muscle cells in inflamed arteries are believed to acquire pro-inflammatory properties and produce several growth factors (vascular endothelial growth factor (VEGF) and platelet-derived growth factor (PDGF), among others), causing intimal hyperplasia and vascular occlusion.

B cells are not present in the arterial wall of all the patients with GCA, suggesting that its effect is not crucial in the pathogenesis. However, when present, they are activated and contribute for the amplification of the immune response and inflammation.

Finally, defects in immune checkpoints have also been studied and appear to play a role in the immune activation observed in GCA. T cells express programmed death-1 (PD-1), which binds to its receptor in antigen-presenting cells, inducing T cell anergy and apoptosis, and the production of IL-10 by T cells or their polarization into T-reg lymphocytes. A defect in the immunoprotective PD-1/PD-L1 immune checkpoint has been reported in GCA patients.

3. Clinical features

Headache is the most common symptom of GCA, and a new-onset headache (or worsening of a preexisting headache) in older adults should always raise the suspicion of this condition. The installation of the pain is usually acute/subacute. The typical pattern is a temporal headache, continuous throughout the day and resistant to standard analgesia, but it can also be felt over other cranial areas or be diffuse. Patients may also present with scalp tenderness or tongue/jaw pain. Jaw claudication, seen in 45% of the patients, is a relatively specific sign and correlates with temporal artery biopsy positivity.

Total or partial visual loss affects 15–20% of patients, mostly at disease onset. The main cause is anterior ischemic optic neuropathy due to vasculitis of the posterior ciliary arteries, which are branches of the ophthalmic artery and responsible for the blood supply of the optic nerve and the choroid. The optic neuropathy leads to visual loss, which is usually painless and with rapid onset. Occasionally, posterior ischemic optic neuropathy, occlusive vasculopathy of the central artery of the retina, or cortical ischemia can cause visual loss too. Blindness is irreversible in most cases, and when one eye is affected, the other one will likely (in half of the cases) be diseased too in a few days if the treatment is not promptly started. About 10% of patients develop amaurosis fugax, visual hallucinations, or diplopia, which are considered premonitory signs and can progress to blindness in half of the cases.

Systemic manifestations, such as fever, weight loss, and fatigue, are frequent (42, 50, and 40%, respectively). Some patients present only with constitutional symptoms, without headache or visual changes, and in those cases, the diagnosis can be challenging. GCA can be the cause of an obscure-origin fever in older patients.

Polymyalgia rheumatica (PMR) is the most common extracranial manifestation of GCA, and it may also be the first clinical sign of a vasculitis relapse. It is defined by pain/tenderness in proximal arms and legs, with morning stiffness and fast improvement with low-dose glucocorticoid treatment. PMR is seen in 50% of the patients with GCA, while only 20% of the patients with PMR have clinical signs of GCA. However, in patients with PMR with persistently high inflammatory markers and insufficient response to glucocorticoids, a careful investigation may reveal an oligo-symptomatic GCA requiring more aggressive treatment. GCA and PMR are different conditions that share many common features: they occur almost exclusively in patients aged 50 years or older, have similar gender ratios, are associated with the HLA-DRB1*04 alleles and increased levels of serum acute-phase reactants, and respond to glucocorticoid therapy.

GCA is a systemic vasculitis with predominance for the aorta and its branches. Therefore, the patients may also present with chest pain and limb claudication. Large-vessel disease may complicate with aneurysms or stenosis development, resulting in increased mortality due to cardiovascular disease. Many patients with GCA develop clinical manifestations of large-vessel involvement, such as arm claudication (4%) and arterial bruits (21%).

Cerebrovascular events are less common ischemic complications, seen in 15% of the patients. They are mostly due to occlusive vasculitis of the carotid or vertebrobasilar arteries. Transitory or persistent ischemic attacks are the most common neurologic manifestations, but cognitive impairment and neuropathy are other possible complications. Neuropsychiatric symptoms (dementia, mood disorders, and psychosis) affect 3% of the patients. Audio vestibular dysfunction leading to sensorineural hearing loss has also been described in GCA.

Infrequent clinical manifestations include tongue, scalp or lip necrosis, and facial/submandibular swelling. Unlike visual loss, ischemic necrosis tends to improve after the glucocorticoid treatment is started. Peripheral synovitis is found in 15% of the patients.

Physical examination is frequently normal, but it may also reveal temporal artery abnormalities such as thickness, tenderness, and hyperemia, with normal, decreased, or absent pulse. Even though temporal artery abnormality is one of the five ACR classification criteria for GCA, it is seen in less than 30% of the patients. Peripheral pulses in arms and legs may be decreased or absent as well in large-vessel disease, and arterial bruits can be heard in such patients.

4. Diagnosis

The American College of Rheumatology classification criteria for GCA, published in 1990, requires three or more of the following five criteria [1]: age 50 years and older [2], new onset of localized headache [3], temporal artery tenderness on palpation or decreased pulsation [4], an abnormal temporal artery biopsy, and [5] an erythrocyte sedimentation rate (ESR) of 50 mm/h or more (**Table 1**) [7]. These criteria are designed to be classificatory and not diagnostic, with sensitivity and specificity of 81.1% and 64.2%, respectively [8]. In the last years, new imaging techniques have emerged and can be helpful tools on diagnosis and disease activity assessment.

1. Age ≥ 50 years	1
2. New-onset headache	1
3. Temporal artery tenderness/decreased pulsation	1
4. Abnormal temporal artery biopsy	1
5. ESR ≥ 50 mm/1st hour	1
Total	
GCA defined if score ≥ 3.	

Table 1.
ACR classification criteria for GCA (1990).

4.1 Acute-phase reactants

ESR higher than 50 mm/h is one of the five criteria for GCA classification according to the ACR. However, in recent years alternative acute-phase reactants have been proposed as more sensitive markers. Recently one study analyzed 26 markers in GCA and PMR comparing with healthy controls and found that three serum markers (B cell-activating factor [BAFF], CXCL9, and IL-6) were increased in both newly diagnosed GCA and newly diagnosed PMR patients. Serum BAFF and IL-6, but not CXCL9, were attenuated upon glucocorticoid-induced remission and showed the strongest association with disease activity in both GCA and PMR patients [9]. BAFF is an important regulator of B cell responses and has been linked to the development of many autoimmune diseases. Even though these markers are not used routinely in clinical practice yet, they are promising new tools in the diagnostic approach and disease activity assessment in GCA.

Platelets are also considered a serum marker for inflammation in GCA. The postulated mechanism of thrombocytosis in promoting inflammation stems from their early interaction with the endothelium in inflammatory states during which they provide adhesion molecules and chemotactic stimulation to aid in the recruitment of leukocytes and enhance the release of different inflammatory mediators [10].

The most used tests for measuring inflammation in clinical practice are ESR and CRP. One study showed that the optimal cutoff value for CRP in GCA was 26.9 mg/L (sensitivity 75% and specificity 51%) and for ESR was 53 mm/h (sensitivity 66% and specificity 55%) [11].

4.2 Temporal artery biopsy (TAB)

Even though temporal artery biopsy is not essential for the diagnosis, one study found that sensitivity and specificity of ACR criteria for diagnosis of GCA before performing TAB were 68.5 and 58%, respectively, while sensitivity and specificity of ACR criteria after performing TAB biopsy were 89.8 and 64.5%, respectively [12]. The sensitivity rates are lower in the large-vessel phenotype of GCA, and even in patients with the temporal artery affected, skip lesions may contribute to a negative TAB.

TAB showing transmural inflammation is still considered the gold standard for the diagnosis of GCA and remains the most specific diagnostic test. It is a minimally invasive and well-tolerated surgical procedure that is generally performed in an outpatient surgery setting and carries a low risk of complications in experienced hands. The length of the artery needed for optimal histopathological analysis is 2 cm. For all these reasons, TAB is still recommended in all patients with a clinical suspicion of GCA. In addition to their diagnostic role, histological findings of positive TAB may have clinical and prognostic significance and thus implications for the patients' management.

GCA is characterized histopathologically by mononuclear infiltrates in all layers of the arterial wall. Macrophages and T cells are present in granuloma formation, and multinucleated giant cells are localized close to the fragmented internal elastic lamina. Neutrophils, eosinophils, and plasma cells are rare. Proliferation of the intima results in occlusive vasculopathy. Neoangiogenesis is frequent and at times prominent, and fibrinoid necrosis is typically absent (**Figures 1** and **2**) [13].

The positivity of TAB declines after glucocorticoid treatment is started, and the biopsy should ideally be performed within 2 weeks from the onset of the therapy.

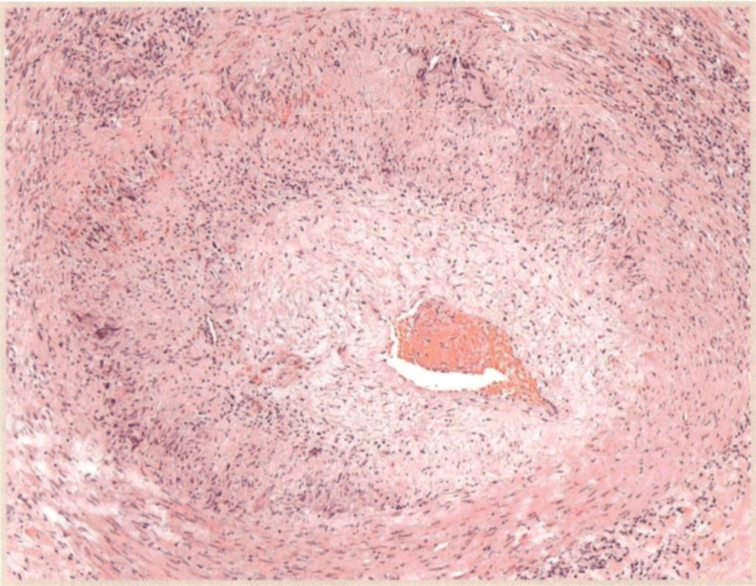

Figure 1.
Histopathological analysis of a patient from the Rheumatology Division of the University of Sao Paulo showing a transmural lymphomonocytary infiltrate and important narrowing of the vessel lumen. HE 100×. Image gently provided by the Pathological Anatomy Division of the University of Sao Paulo, School of Medicine.

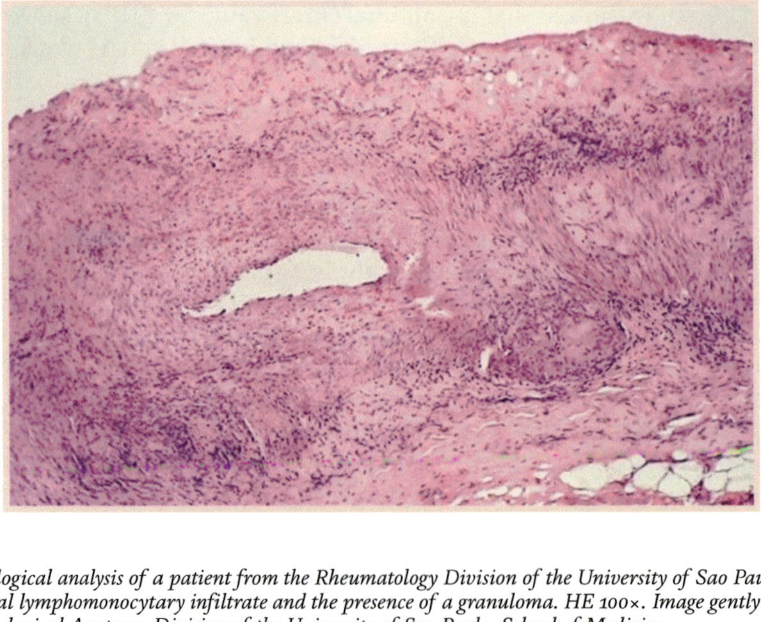

Figure 2.
Histopathological analysis of a patient from the Rheumatology Division of the University of Sao Paulo showing a transmural lymphomonocytary infiltrate and the presence of a granuloma. HE 100×. Image gently provided by the Pathological Anatomy Division of the University of Sao Paulo, School of Medicine.

Importantly, the therapy should never be delayed for the performance of a biopsy, especially when visual symptoms are present [14].

For patients with a negative temporal artery biopsy, clinical assessment remains a mainstay of diagnosis.

4.3 Imaging techniques

In 2018, the European League Against Rheumatism (EULAR) published recommendations about the use of imaging techniques in large-vessel vasculitis (LVV), which included GCA and Takayasu's arteritis. According to these recommendations, in patients with suspected GCA, an early imaging test is recommended to complement the clinical criteria for diagnosing GCA, assuming high expertise and prompt availability of the imaging technique [15]. However, imaging should not delay initiation of treatment. The choice of the individual imaging method depends on the predominant clinical symptoms and local settings. In settings where imaging modalities are not readily available or expertise with imaging in GCA is questionable, a biopsy should still be favored in first place. Besides, if positive histology is already available, additional imaging may not be needed for the diagnosis. In centers, however, where imaging (and TAB) is readily available and performed with high quality, the task force recommends that imaging should be preferred as the first test because of low invasiveness, ready availability of imaging results, and assessment of a larger extent of potentially inflamed arteries at the same examination, therefore contributing to a lower number of false-negative results. Imaging should be performed before or as early as possible after initiation of therapy, best within 1 week, because treatment with glucocorticoids rapidly reduces the sensitivity of imaging [15].

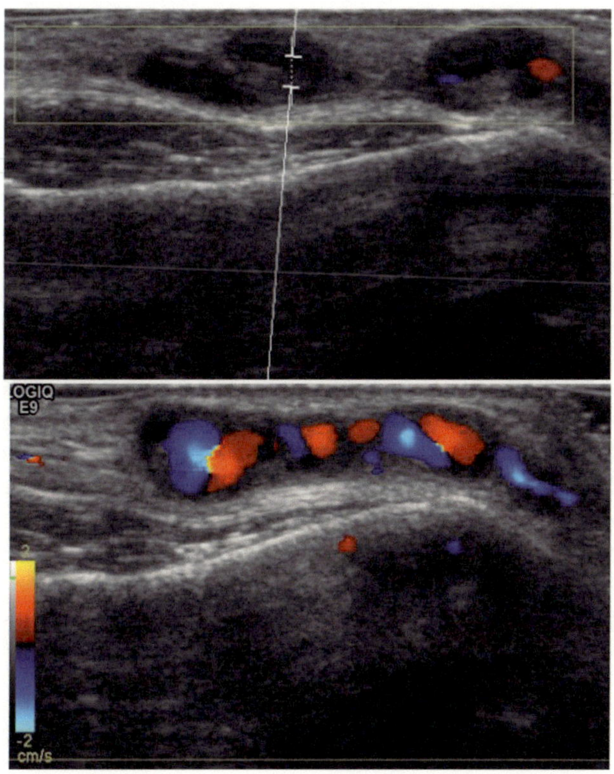

Figure 3.
Temporal artery duplex scan of a patient from the Rheumatology Division of the University of Sao Paulo, School of Medicine with GCA showing thickness in the vascular wall and the noncompressible "halo" sign.

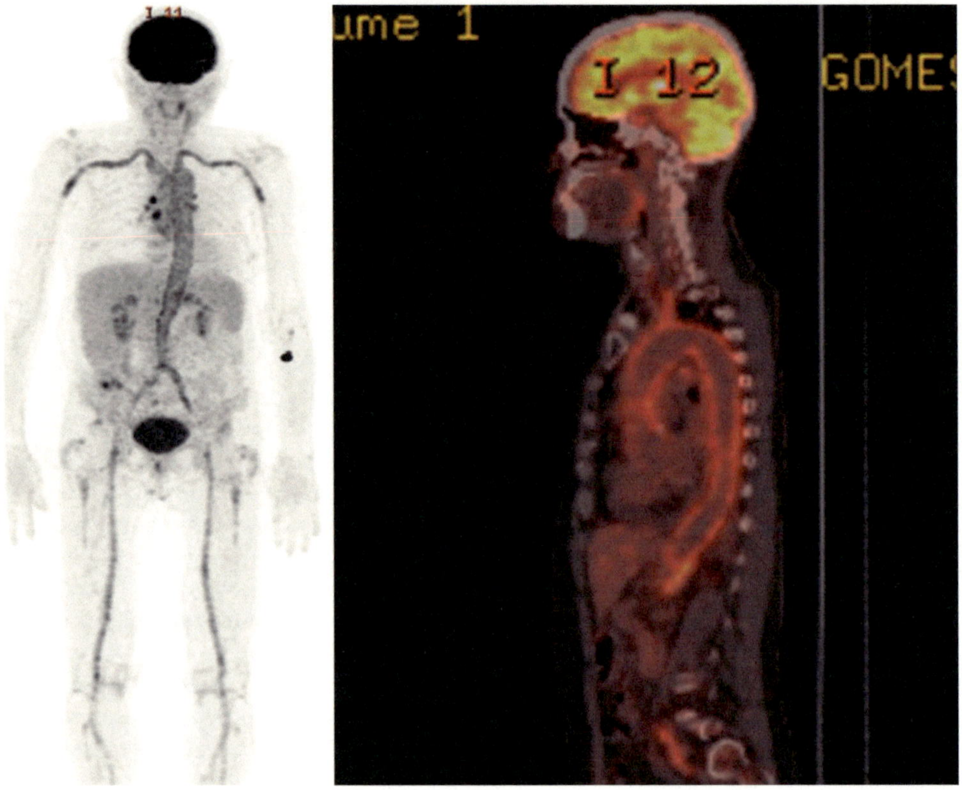

Figure 4.
PET-CT of a patient from the Rheumatology Division of the University of Sao Paulo, School of Medicine with GCA showing inflammation in the vascular wall of the aorta and subclavian, common carotid, iliac, femoral, popliteal, and tibial arteries.

In patients in whom there is a high clinical suspicion of GCA and a positive imaging test, the diagnosis of GCA may be made without an additional test (biopsy or further imaging). In patients with a low clinical probability and a negative imaging result, the diagnosis of GCA can be considered unlikely [15].

Ultrasound of temporal ± axillary arteries is recommended as the first imaging modality in patients with suspected predominantly cranial GCA. A noncompressible "halo" sign is the ultrasound finding most suggestive of GCA (**Figure 3**). High-resolution MRI of cranial arteries to investigate mural inflammation may be used as an alternative for GCA diagnosis if ultrasound is not available or inconclusive. Ultrasound, PET, MRI, and/or CT may be used for the detection of mural inflammation and/or luminal changes in extracranial arteries to support the diagnosis of large-vessel GCA (**Figure 4**). Ultrasound is of limited value for the assessment of aortitis [15].

In patients with a suspected flare, imaging might be helpful to confirm or exclude it. Imaging is not routinely recommended for patients in clinical and biochemical remission. In patients with large-vessel vasculitis, MRA, CTA, and/or ultrasound may be used for long-term monitoring of structural damage, particularly to detect stenosis, occlusion, dilatation, and/or aneurysms [15].

5. Treatment

Treatment with oral glucocorticoid (GC) effectively induces remission and reduces the evolution to visual loss, and it should be started as early as possible when there is a clinical suspicion of GCA. The GC therapy cannot be postponed to after confirmation of the diagnosis, because once the visual loss is installed, it is rarely reversible [16].

Oral prednisone in a daily single dose of 40–60 mg usually resolves the symptoms and normalizes acute inflammation reactants within the first 2–4 weeks of the treatment. When premonitory visual signs are present (amaurosis fugax) or when visual loss is installed, pulse therapy with daily intravenous methylprednisolone (500–1000 mg) for 3 days can be tried, even though its superiority compared to the oral prednisone regimen has not been proven in clinical trials.

Glucocorticoids are effective in inducing clinical remission, but the side effects of its chronical use are undesirable, especially in elderly individuals. Therefore, synthetic or biological immunosuppressants have been used as GC-sparing adjuvants to reduce the cumulative GC dose and to maintain remission after the prednisone withdrawal [17]. There is no consensus on the timing of initiating GC-sparing therapy, but indications to start it early in the disease course include the presence of significant premorbid diseases (diabetes mellitus, osteoporosis, obesity), the emergence of significant glucocorticoid-related side effects, and a relapsing course necessitating protracted CS use. After clinical remission is achieved (symptoms resolved and laboratory inflammation markers normalized), the GC taper can be started. It has to be slow, especially with lower doses. The dose can gradually be reduced by 5 mg every 2 weeks to 20 mg/day and then by 2.5 mg every 2 weeks to 10 mg/day if there are no flares of disease activity. After achieving a daily dose of 10 mg, the prednisone taper should be slowed, such that patients remain on progressively decreasing doses over the ensuing 6–12 months. Tapering by 1 mg decrements each month once the daily dose is less than 10 mg can be considered. Disease relapses are more frequent in this final phase of the GC tapering regimen [18].

5.1 Glucocorticoid-sparing therapy

Methotrexate (MTX) is the conventional immunosuppressive drug most commonly used for the management of refractory GCA. However, the efficacy of this drug in GCA is modest. The trials yielded a role of MTX (10–15 mg/week) to reduce the frequency of relapses (by 35% of a first relapse and by 51% of a second relapse) and decrease the cumulative prednisone dose. However, the optimum efficacy of MTX becomes manifest only after 24–36 weeks [19].

Leflunomide may also be an effective and well-tolerated glucocorticoid-sparing agent in GCA, but there are no randomized controlled trials to confirm it yet. In one prospective observational study with 76 newly diagnosed GCA patients, 10 mg daily leflunomide was compared with glucocorticoid only in a follow-up period of at least 48 weeks. During the follow-up 13.3% patients in the leflunomide group flared versus 39.1% in the GC-only group ($p = 0.02$). Furthermore, 56.7% patients in the leflunomide were able to stop GC at week 48 but none in the GC-only group [20].

Tocilizumab is a humanized monoclonal antibody that binds to the soluble and membrane-bound forms of the IL-6 receptor (IL-6R). IL-6 has a key role in the pathogenesis of GCA, and elevated levels of IL-6 are present and correlate with disease activity. Efficacy of tocilizumab in GCA has been proved in a multicenter, randomized, double-blind, placebo-controlled, phase 3 trial with 251 patients (119 newly diagnosed and 132 with relapsing disease). The patients were randomized to receive subcutaneous tocilizumab (162 mg) weekly or every other week combined with a 26-week prednisone taper or placebo, combined with a prednisone taper over a period of either 26 or 52 weeks. Both groups of patients treated with tocilizumab achieved sustained remission more commonly than those placebo-treated at week 52. Patients who underwent tocilizumab therapy had fewer relapses of disease than those in the placebo arms [21]. Tocilizumab use was associated with a powerful glucocorticoid-sparing effect, and this effect was stronger in those patients who had experienced relapses before randomization. Tocilizumab in a subcutaneous weekly dose of 162 mg was approved by the US FDA and the European Commission for the treatment of GCA. However, there is concern about the characterization of a relapse in patients that are under tocilizumab therapy, because tocilizumab is very effective in normalizing CRP and ESR and some patients can be oligo-symptomatic during a flare. Periodic imaging in these patients is recommended for accessing vascular activity or progression of the vascular damage during the treatment.

Ustekinumab is a human monoclonal antibody that binds to the p40 subunit of both IL-12 and IL-23 preventing their binding to their shared cell surface receptor chain, IL-12β. The inhibition of IL-12 signaling abrogates Th1 response with reduction in TNF-α, IFN-γ, and IL-12 production. The inhibition of IL-23 signaling abrogates Th17 response with the reduction on IL-6, IL-17, IL-21, IL-22, and TNF-α production. Th1 and Th17 responses both have important roles in the pathogenesis of GCA. A prospective open-label 52-week trial with 25 patients with relapsing GCA showed that ustekinumab may be effective for the treatment of GCA: at week 52 the median daily dose of prednisolone decreased from 20 to 5 mg ($p < 0.0001$), and no patient experienced a relapse of GCA while receiving ustekinumab [22]. No randomized controlled trial with ustekinumab in GCA patients has been performed yet.

Abatacept is a fully human fusion protein that binds to CD80/CD86 on antigen-presenting cells preventing these molecules from binding to their ligand, CD28, on T cells, and is moderately effective in the treatment of GCA. In a multicenter, randomized, double-blind trial, 49 patients with GCA were treated with 10 mg/kg intravenous abatacept on days 1, 4, and 29 and week 8 and monthly after that. The relapse-free survival rate at 12 months, the primary endpoint, was 48% for those receiving abatacept and 31% for those receiving placebo ($p = 0.049$) [23]. Further studies encompassing larger number of patients are needed to confirm the utility of abatacept as adjunctive therapy in GCA.

Anti-TNFα agents have been tested and yielded disappointing results, showing no efficacy in reducing GC dose or relapse rates in GCA. There are other promising target therapies being tested, such as the JAK/STAT inhibitors, but the results are not available, and there is no data to support their use in clinical practice yet.

5.2 Preventing complications

Patients with GCA are elderly and frequently have multiple comorbid conditions that can be worsened by the use of GC and immunosuppressants. Therefore, the levels of vitamin D must be higher than 30 ng/mL for all patients, and the dietary intake of calcium must be stimulated (or supplementation, if the dietary intake is insufficient) for bone protection, as well as the use of bisphosphonates if indicated.

Low-dose aspirin (80–100 mg/day) should be prescribed for prevention of cardiovascular events, which represent the main cause of death in this population.

6. Prognosis

The most frequent causes of death in GCA patients are cardiovascular diseases followed by cancer. Combined, these conditions account for approximately two thirds of all deaths. A Norwegian cohort of 881 patients with GCA and 2577 population controls found no significant difference in the overall cumulative survival or survival at any specific time after diagnosis. In this study the mean age of death was

83.6 (SD 7.5) years for GCA patients, and survival was more than 80% in 5 years and approximately 50% in 10 years [24]. The same study found that even though the overall mortality was not reduced in GCA, these patients have an increased risk of death due to circulatory diseases and infections but a decreased risk of death due to cancer over time. The increased risk of death by circulatory diseases may be related to aneurysms and dissections, which are recognized as large-vessel complications of GCA. Therefore, it is extremely important in the management of these patients to identify and to treat other contributing risk factors for circulatory disease.

Author details

Marília A. Dagostin and Rosa M.R. Pereira*
Rheumatology Division, Hospital das Clínicas HCFMUSP, Faculdade de Medicina, Universidade de Sao Paulo, Sao Paulo, SP, Brazil

*Address all correspondence to: rosamariarp@yahoo.com

References

[1] Hutchinson J. Disease of the arteries: On a peculiar form of thrombotic arteritis of the aged which is sometimes productive of gangrene. Archives of Surgery. 1890;**1**:323-329

[2] Horton BT, Magath BT, Brown GE. Arteritis of temporal vessels: Report of 7 cases. Proceedings of the Staff Meetings. Mayo Clinic. 1937;**12**:548-553

[3] Jennette JC et al. 2012 revised international Chapel Hill Consensus Conference Nomenclature of Vasculitides. Arthritis and Rheumatism. 2013;**65**(1):1-11

[4] Hochberg M, Gravallese E, Silman A, Smolen J, Weinblatt M, Weisnman M. Rheumatology. 7th ed. Philadelphia, PA, USA: Elsevier; 2018

[5] Mackie SL, Taylor JC, Haroon-Rashid L, Martin S, Dasgupta B, Gough A, et al. Association of HLA-DRB1 amino acid residues with giant cell arteritis: Genetic association study, meta-analysis and geo-epidemiological investigation. Arthritis Research & Therapy. 2015;**17**:195

[6] Espígol-Frigolé G, Planas-Rigol E, Lozano E, Corbera-Bellalta M, Terrades-García N, Prieto-González S, et al. Expression and function of IL12/23 related cytokine subunits (p35, p40, and p19) in giant-cell arteritis lesions: Contribution of p40 to Th1-and Th17-mediated inflammatory pathways. Frontiers in Immunology. 2018;**9**:809

[7] Bloch DA, Michel BA, Hunder GG, McShane DJ, Arend WP, Calabrese LH, et al. The American College of Rheumatology 1990 criteria for the classification of vasculitis. Arthritis and Rheumatism. 1990;**33**:8

[8] Seeliger B, Sznajd J, Robson JC, Judge A, Craven A, Grayson PC, et al. Are the 1990 American College of Rheumatology vasculitis classification criteria still valid? Rheumatology (Oxford). 2017;**56**(7):1154-1161

[9] Van der Geest KSM, Abdulahad WH, Rutgers A, Horst G, Bijzet J, Arends S, et al. Serum markers associated with disease activity in giant cell arteritis and polymyalgia rheumatica. Rheumatology. 2015;**54**:1397-1402

[10] Chan FLY, Lester S, Whittle S, Hill CL. The utility of ESR, CRP and platelets in the diagnosis of GCA. BMC Rheumatology. 2019;**3**:14

[11] Kermani TA, Schmidt J, Crowson CS, Ytterberg SR, Hunder GG, Matteson EL, et al. Utility of erythrocyte sedimentation rate and C-reactive protein for the diagnosis of giant cell arteritis. Seminars in Arthritis and Rheumatism. 2012;**41**:866-871

[12] Grossman C, Barshack I, Bornstein G, Ben-Zvi I. Is temporal artery biopsy essential in all cases of suspected giant cell arteritis? Clinical and Experimental Rheumatology. 2015;**33**(2 Suppl 89):S-84-S-89

[13] Buttgereit F, Dejaco C, Matteson EL, Dasgupta B. Polymyalgia rheumatica and giant cell arteritis. JAMA. 2016;**315**(22):2442

[14] Ing EB et al. Systematic review of the yield of temporal artery biopsy for suspected giant cell arteritis. Neuro-Ophthalmology. 2019;**43**(1):18-25

[15] Dejaco C et al. EULAR recommendations for the use of imaging in large vessel vasculitis in clinical practice. Annals of the Rheumatic Diseases. 2018;**77**:636-643

[16] Low C, Conway R. Current advances in the treatment of giant cell arteritis: The role of biologics. Therapeutic

Advances in Musculoskeletal Disease. 2019;**11**:1-11

[17] Winkler A, True D. Giant cell arteritis: 2018 review. Missouri Medicine. September/October 2018;**115**:5

[18] Salvarani C, Hatemi G. Management of large-vessel vasculitides. Current Opinion in Rheumatology. 2019;**31**:25-31

[19] Mahr AD, Jover JA, Spiera RF, Hernández-García C, Fernández-Gutiérrez B, La Valley MP, et al. Adjunctive methotrexate for treatment of giant cell arteritis—An individual patient data meta-analysis. Arthritis and Rheumatism. 2007;**56**(8):2789-2797

[20] Hocevar A, Ješe R, Rotar Ž, Tomšič M. The role of leflunomide in the treatment of giant cell arteritis. Annals of the Rheumatic Diseases. 2018;**77**(Suppl 2):1114-1115

[21] Stone JH, Tuckwell K, Dimonaco S, Klearman M, Aringer M, Blockmans D, et al. Trial of tocilizumab in giant-cell arteritis. The New England Journal of Medicine. 2017;**377**(4):317-328

[22] Conway R, O'Neil L, Gallagher P, McCarthy GM, Murphy CC, Veale DJ, et al. Ustekinumab for refractory giant cell arteritis: A prospective 52-week trial. Seminars in Arthritis and Rheumatism. 2018;**48**:523-528

[23] Langford CA, Cuthbertson D, Ytterberg SR, Khalidi N, Moach PA, Carette S, et al. A randomized, double-blind trial of abatacept (CTLA4-IG) for the treatment of giant cell arteritis. Arthritis & Rheumatology. 2017;**69**:837-845

[24] Brekke LK, Fevang BTS, Diamantopoulos AP, Assmus J, Esperø E, Gjesdal CG. Survival and death causes of patients with giant cell arteritis in Western Norway 1972-2012: A retrospective cohort study. Arthritis Research & Therapy. 2019;**21**:154

Chapter 7

Anomalous Origin of Coronary Arteries

Xhevdet Krasniqi and Hajdin Çitaku

Abstract

Coronary arteries supply the heart muscle with blood maintaining myocardial hemostasis and function. Coronary artery anomalies may persist after birth affecting cardiovascular system through haemodynamic impairment caused from shunting, ischaemia, especially in young children or adolescents and young adults. In patients undergoing coronary angiography the incidence of anomalous origination of the left coronary artery from right sinus is 0.15% and the right coronary artery from the left sinus is 0.92%. A recent classification of the coronary anomalies is based on anatomical considerations, recognizing three categories: anomalies of the origin and course, anomalies of the intrinsic coronary artery anatomy, and anomalies of the termination. In the setting of anomalous coronary artery from the opposite sinus, the proximal anomalous CA may run anterior to the pulmonary trunk (prepulmonic), posterior to the aorta (retroaortic), septal (subpulmonic), or between the pulmonary artery and the aorta itself (interarterial). Among them, only those with an interarterial aorta-pulmonary course are regarded as hidden conditions at risk of ischaemia and even sudden death. We presented two cases with anomalous origin of coronary arteries from opposite sinus, and two other cases with anomalous origin of left circumflex artery. The atherosclerotic coronary artery disease leads to the need of coronarography which can find out the presence of coronary artery anomalies. Anomalous origin of coronary artery that is present with atherosclerotic changes continues to exist as a challenge during treatment in interventional cardiology.

Keywords: coronary arteries, anomalous origin, opposite sinus

1. Introduction

Coronary arteries (CAs) are the blood vessels that supply the heart muscle with blood.

Intact coronary circulation is therefore important for myocardial hemostasis and function, thus enabling rest of body to function. The disruption of coronary development during embryogenesis results in coronary congenital defects such as coronary mispatterning, structural vascular defects, and anomalous communication of coronary vessels, they can alter coronary artery blood flow.

Such anomalies may persist after birth, occasionally they are in association with other cardiac conditions, so can severely affect cardiovascular system through haemodynamic impairment caused from shunting, ischaemia, or even sudden cardiac death, especially in young children or adolescents and young adults.

The estimated prevalence of CA anomalies is not quite clear with variable, ranging from 0.21 to 5.79% based on angiography, computed tomography (CT), and autopsy databanks [1]. Congenital anomalies of coronary arteries have in incidence about 1% in patients undergoing coronary angiography while the incidence of anomalous origination of the left coronary artery from right sinus is 0.15% and the right coronary artery from the left sinus is 0.92% [2, 3].

2. Coronary artery anatomy

In normal anatomy, the LAD and Cx originate from an aortic area located above the upper or middle third of the left coronary sinus of Valsalva (also called the left posterior sinus). The right coronary artery originates from the upper or middle third of the right sinus of the Valsalva. Normally, the coronary ostia lead an orthogonally oriented coronary proximal stem, off the aortic wall.

Of the many coronary arteries, the "primary" (or elementary) ones are defined as the three main proximal arteries: one provide circulation to the anterior septum and anterior lateral wall (the left anterior descending or LAD), another provides blood flow to the obtuse marginal region of the left ventricle (the circumflex, or Cx), and the third provides circulation to the free wall of the right ventricle (the right coronary artery or RCA). The left main trunk may serve as a common stem that joins the LAD and Cx (a common left main stem is present in about 90% of the cases and is not essential, but the LAD and CX are essential). Normally, the LAD and Cx originate from an aortic area located above the upper or middle third of the left coronary sinus of Valsalva (also called the left posterior sinus) [4]. In **Figure 1** we presented normal origin of coronary arteries.

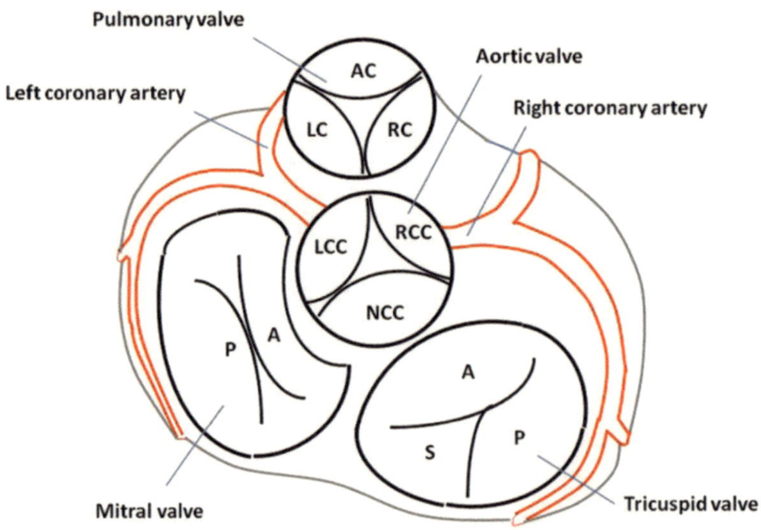

Figure 1.
Normal origin of coronary artery.

3. Embryology of coronary artery

Coronary artery formation is a process involving vasculogenesis, angiogenesis, and arteriogenesis. The vasculogenesis is a process through which is formed the early arterial coronary vascular system via the coalescence of the endothelial precursor cells (angioblasts), and subsequent fusion of the endothelial cell clusters [5].

Angiogenesis implies the generation of the new microvessels by endothelial proliferation and migration, mostly by means of controlled endothelial sprouting [6]. Coronary artery (CA) were originally thought to form bay angiogenesis from the aortic root endothelium based on anatomical facts that ones join the systemic circulation at the aortic root, whereas cardiac veins connect to the general circulation via the coronary sinus [7].

Arteriogenesis describes the remodeling that form mature arteries by migration of supporting smooth muscle cells (SMCs) and pericytes from the epicardium during development [8, 9].

4. Congenital anomalies of the coronary circulation

Coronary anomalies are defined as those angiographic findings in which the number, origin, course and termination of the arteries are rarely encountered in general population. Coronary anomalies may occur in 1–5% of the patients undergoing coronary arteriography, depending on the threshold for defining an anatomic variant [10–12].

A recent classification of the coronary anomalies (**Table 1**) [1] is based on anatomical considerations, recognizing three categories:

a. anomalies of the origin and course;

b. anomalies of the intrinsic CA anatomy; and

c. anomalies of the termination [13, 14].

I. Anomalies of origin coronary artery connection
1. Anomalous origin to the aorta
a. Absent of left main trunk
b. Anomalous coronary artery ostium location
c. Anomalous coronary artery location at improper aortic sinus-wrong sinus:
RCA to left sinus
LCA to right sinus
LCX to RCA/or sinus
RCA or LCA to posterior sinus
with anomalous course: interarterial, prepulmonic, intraseptal, retroaortic, posterior atrioventricular groove or retrocardiac, postero-anterior interventricular groove
d. Single coronary artery
e. Anomalous coronary artery ostium location outside sinu-tubular aorta:
LV
Ascending aorta
Aortic arch
Others (innominate artery; right carotid artery; internal mammary artery; bronchial artery; subclavian artery; descending thoracic aorta)

2. Anomalous origin to the pulmonary artery
a. LCA to posterior facing sinus (ALCAPA)
b. LCX to posterior facing sinus
c. LAD to posterior facing sinus
d. RCA to anterior right facing sinus
e. Ectopic connection (outside facing sinuses) of any CA to PA left sinus, trunk, or branch
f. RV
II. Anomalous intrinsic coronary artery anatomy
1. Congenital ostial stenosis or atresia (LCA, LAD, RCA, Cx)
2. Coronary ostial dimple
3. Coronary ectasia or aneurysm
4. Absent coronary artery
5. Coronary hypoplasia
6. Intramural coronary artery (myocardial bridge)
7. Subendocardial coronary course
8. Coronary crossing
9. Anomalous origination of posterior descending branch or septal penetrating branch
10. Absent PD or split RCA
11. Absent or split LAD
12. Ectopic origination of first septal branch
III. Coronary artery interaction
1. Inadequate arteriolar/capillary ramifications
2. Fistulae from RCA, LCA, or infundibular artery to: RV, RA, coronary sinus, superior vena cava, PA, PV, LA, LV, multiple

Table 1.
Classification of coronary artery anomalies.

4.1 Anomalous pulmonary origin of the coronary arteries (APOCA)

This syndrome is characterized by the origin of the coronary artery arising from the pulmonary artery. The most variant is an anomalous origin of the LCA from the pulmonary artery. (ALCAPA) [15, 16] although single-vessel origins of the RCA, LCx coronary, or LAD artery from the pulmonary artery have also been reported. If untreated, and in the absence of an adequate collateral network, most (95%) infants with APOCA will die within the first year. In the presence of an extensive collateral network, patients may survive into adulthood.

Aortography reveals a large RCA with absence of a left coronary ostium in the left aortic sinus of Valsalva, and with LAD and Cx branches filling through collateral circulation from the RCA branches. Still very delayed in filming sequence retrograde flow from LAD and LCx opacifies the LMCA and its origin from the main pulmonary artery. Still later in the filming sequence retrograde flow from the LAD and LCx arteries opacifies the LMCA and its origin from the main pulmonary artery. Once it is diagnosed, CABG surgery is recommended because of the high incidence of sudden death, cardiomyopathy and arrhythmias associated with APOCA.

4.2 Anomalous coronary artery from the opposite sinus (ACAOS)

Anomalous origin of either the RCA to the left coronary sinus or the LCA to the right coronary sinus, the proximal anomalous coronary artery (CA) may run anterior to the pulmonary trunk (prepulmonic), posterior to the aorta (retroaortic), septal (subpulmonic), or between the pulmonary artery and the aorta (interarterial). Only those with an interarterial (aorta-pulmonary) course can increase risks of myocardial ischemia, arrhythmia, syncope, and sudden cardiac death considering life threatening and clinical guidelines recommend surgical correction [17].

Numerous mechanisms of ischaemia particularly during exercise have been suggested: (1) the compression of the anomalous vessel coursing between the aorta and the pulmonary artery during increased cardiac output and expansion of the great vessels; (2) the acute angle takeoff of the anomalous vessel with further stretch during exercise, possibility accounting for a flap-like closure of the coronary ostium; (3) spasm or kinking of the anomalous vessels; and (4) the course within the aortic wall ("intramural") of the proximal segment of the anomalous vessel [13, 17]. The intramural aortic course can explain the imaging feature (angiography and echo) of CA intussusceptions into the aoric wall: the proximal segment of the anomalous vessel (segmental hypoplasia) is narrowed, and the asymmetrical lateral compression of the anomalous vessel with a silt-like or ovoid rather than circular lumen, particularly during systole and stress.

In **Figure 2** (A and B) we present anomalous Coronary Artery from the Opposite Sinus (ACAOS).

4.3 Coronary artery fistulas

Coronary artery fistulas are defined as abnormal communications between a coronary artery and a cardiac chamber or major vessel, such as to the vena cava, right or left ventricle, pulmonary vein or pulmonary artery [18, 19]. Coronary artery fistulas are rare findings, identified in 10 (0.05%) of 18,272 diagnostic cardiac catheterizations [20].

4.4 Myocardial bridging

The three major coronary arteries generally course along the epicardial surface of the heart. On occasion, however, short coronary artery segments descend into the myocardium for a variable distance. This abnormality, termed myocardial bridging occurs in 5–12% patients and usually is confined to LAD [21].

5. Clinical presentation

5.1 Anomalous pulmonary origin of the coronary arteries (APOCA)

Coronary steal syndrome results where an alteration of circulation patterns leads to a reduction in the blood directed to the coronary circulation. The low pressure in the pulmonary artery causes blood from the abnormal LCA to flow towards the pulmonary artery instead of towards the heart resulting in ischaemia and collateral growth.

The extent of the acquired circulation between the two CAs is the major determinant of the degree of ischaemia, severity of clinical presentation, and outcome. Depending collateral vessels clinically are presented: a) the adult-type with well-established collateral vessels, and b) the infant-type without or with few collaterals, with early onset of symptoms when pulmonary arterial pressure decreases [22].

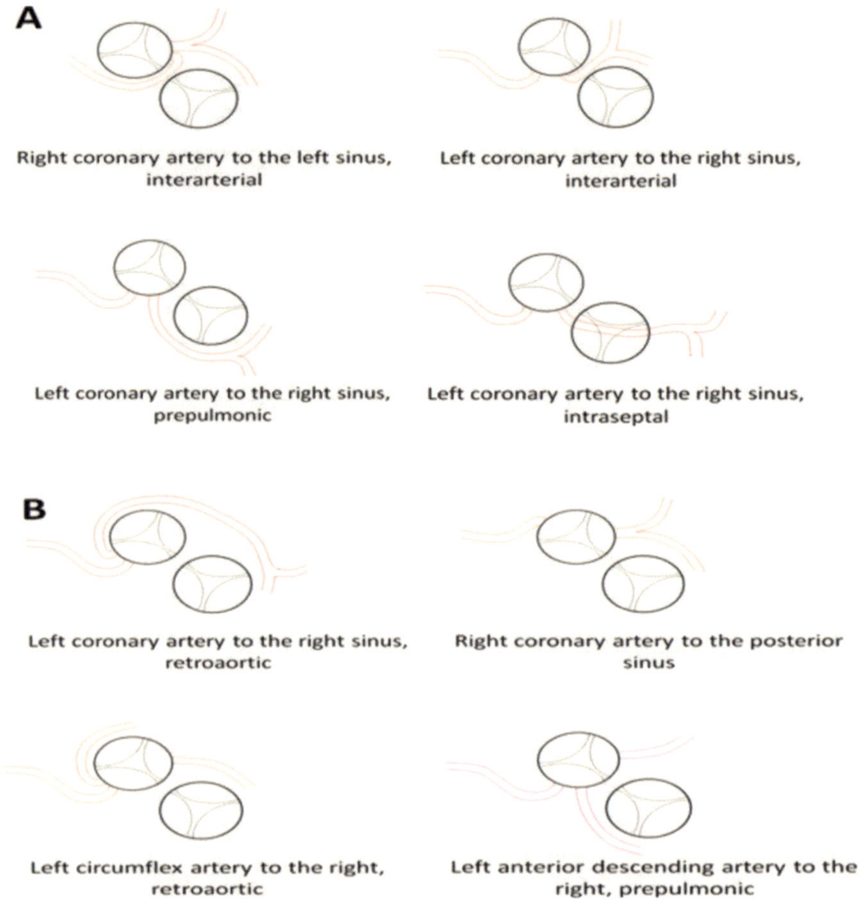

Figure 2.
Anomalous origin of coronary arteries to the aorta.

5.2 Anomalous coronary artery from the opposite sinus (ACAOS)

Origin of LCA from the proximal RCA or the right aortic sinus with subsequent passage between the aorta and the right ventricular outflow tract has been associated with sudden death during or shortly after exercise in young persons.

The increased risk of sudden death may be due to a silt-like ostium, a bend with acute takeoff angles of the aberrant coronary arteries, or arterial compression between the pulmonary trunk and aorta when there is increased blood flow through these vessels with exercise and stress.

The RCA originated from the LCA or left aortic sinus with passage between the aorta and the right ventricular outflow tract is also associated with myocardial ischemia and sudden death [23]. In rare cases of the LCA originated from the right sinus myocardial ischemia may occur even if the LCA passes anterior to the right ventricular outflow tract or posterior to aorta, not through a tunnel between the two great vessels [24].

The revascularization approach in patients with ACAOS has been CABG surgery, although coronary stenting has been reported with acceptable medium term success.

5.3 Coronary artery fistulas

The clinical presentation associated with coronary artery fistulas is dependent on the type of fistula, shunt volume, in situ of the shunt and presence of other cardiac conditions, although patients (50%) often remain asymptomatic [25]. Dyspnea on exertion, fatigue, congestive heart failure, pulmonary hypertension, bacterial endocarditis and arrhythmias are common presentations in symptomatic patients. Myocardial ischaemia may also occur, but the mechanism remains speculative [25]. Symptomatic patients or those with severe shunts may be treated with surgical closure, although percutaneous closure with coil embolization may also be tried.

5.4 Myocardial bridging

A myocardial bridge occurs when one of the coronary arteries takes a tunneled intramuscular course under a bridge of overlying myocardium. The myocardial fibers passes over the involved segment of the LAD, and each contraction of these fibers can cause narrowing of the artery. On angiography, the bridged segment is of normal caliber during diastole and abruptly narrows with each systole.

Although bridging is not thought to have any hemodynamic significance in most cases, myocardial bridging has been associated with angina, arrhythmia, depressed left ventricular function, myocardial stunning, early death after cardiac transplantation, and sudden death [21, 26]. Intracoronary Doppler studies have shown that diastolic flow abnormalities may be present in patients with myocardial bridging. Medical treatment generally includes beta blockers, although nitrates should be avoided because they may worsen symptoms. Intracoronary stent and surgery have been attempted in selected patients, but the results have been mixed.

5.5 Percutaneous coronary intervention and anomalous coronary artery

Several problems may be encountered during the angiography and angioplasty of anomalous origin of culprit coronary artery (AOCCA), including precise diagnosis, selection of an appropriate guiding catheter, insufficient backup force, and difficulties in balloon or stent delivery. The final success of the procedure is depended from the careful assessment of the AOCCA configuration, proximal angulation, vessel course and subsequent selection of an appropriate guide catheter and guide wire.

5.6 Femoral versus radial approach

In case of AOCCA femoral access may offer better options allowing for easy, and multiple catheter exchanges [27]. Although, in the setting of ACS, the operator is usually unaware of AOCCA presence, having to make the best use of the chosen access site. Also, it seems best to use the approach one is most comfortable with as there is usually a way to perform successful PCI of AOCCA regardless of access site.

5.7 Right versus left radial approach

In a meta-analysis of 12 prospective randomized trials comparing above-mentioned approaches there was a small but statistically significant difference in terms of contrast use and fluoroscopy time in favor of coronary procedures performed via left radial approach compared to the right radial approach, but without any difference in access site or other procedural complications [28].

5.8 Additional tools

Anchoring balloons or anchor wire techniques may be helpful tools [29]. The latter maneuver was used to treat one of the present patients. Still, this culprit was not proven for AOCCA. Extension catheters, such as Guideliner or Guidezilla often allow for safe and stable intubation and facilitate stent placement.

6. Case presentations

6.1 Case 1

First case is a 62-year-old female patient hospitalized in our clinic due to chest pain with a history of arterial hypertension and diabetes mellitus. Cardiac biomarkers showed: serum creatinine kinase (CK) level of 82 IU/L, creatinine kinase-myocardial band (CK-MB) level of 33.6 IU/L, and troponin-T level of 684 ug/L. Electrocardiography (ECG) is characterized with ST segment depression in V1–V3. Transthoracic echocardiography (TTE) presented regional wall motion abnormality in the entire severely hypokinetic inferior wall.

The coronary angiography revealed the left coronary artery arising from the right coronary sinus sharing a same ostium with right coronary artery (**Figure 3**). The proximally and distally stenosed left anterior descending artery (LAD) (**Figure 4**) associates with calcified atherosclerotic medial and distal right coronary artery (RCA) stenosis (**Figure 5**).

6.2 Case 2

The second case is a 47-year-old male who presented to emergency department with chest pain. He also had a history of arterial hypertension and a positive history for ischemic heart disease. Cardiac biomarkers: serum creatinine kinase (CK), creatinine kinase-myocardial band (CK-MB), and troponin-T were not increased. Electrocardiography (ECG) showed atypical ST segment changes in leads V4–V6. Transthoracic echocardiography (TTE) did not present regional wall motion abnormalities.

A coronary angiogram showed an anomalous right coronary artery arising from the left Valsalva sinus from a separate ostia with the left coronary artery (**Figures 6** and 7). Medial and distal segments of LAD were tortuous (**Figure 8**).

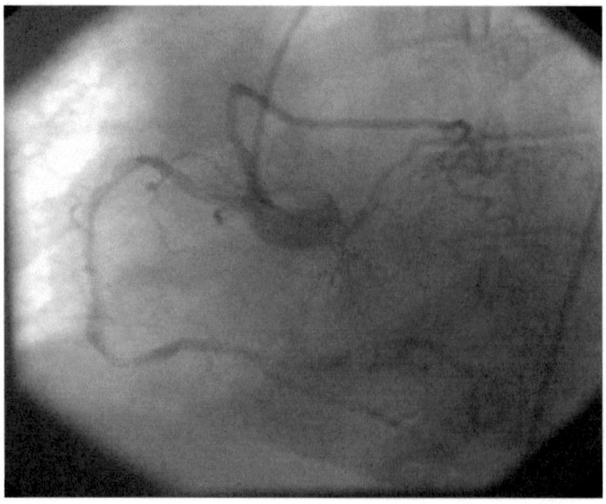

Figure 3.
Coronary angiography revealed a left coronary artery arising from the right Valsalva sinus sharing a same ostium with right coronary artery.

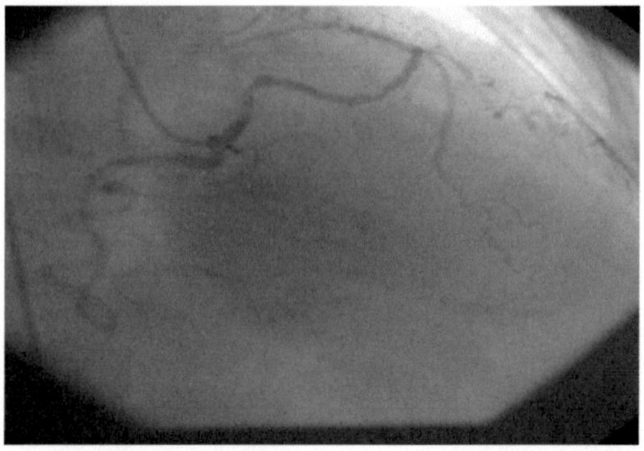

Figure 4.
Stenosis of the proximal and distal segments of LAD.

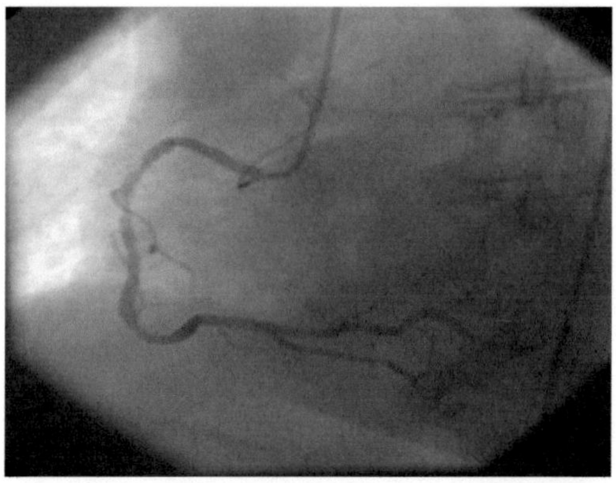

Figure 5.
Selective cannulation of RCA. Calcified atherosclerotic medial and distal RCA stenosis.

6.3 Case 3

The third case is 64-year-old man hospitalized to our clinic due to of chest pain. Also, patient was a smoker and had a history of arterial hypertension, obesity, and dyslipidemia. Biochemical parameters were: serum creatinine kinase (CK) level of 82 IU/L, creatinine kinase-myocardial band (CK-MB) level of 6.5 ng/mL, and troponin-I level of 0.1 ng/mL. Electrocardiography (ECG) is characterized with deep Q wave in inferior and V4–V6 leads with biphasic T in inferior and V3–V6 leads. In transthoracic echocardiography (TTE) is presented with regional motion abnormalities in the entire severely hypokinetic inferoposterio wall.

The patients underwent coronary angiography that revealed the LCx arising from the right coronary sinus (**Figure 9**). The mildly stenosed LCx coexists with a stenosed RCA.

6.4 Case 4

The second case is 67-year-old man presented to the emergency department with chest pain that had developed 6 h previously. That patient had a history of arterial hypertension for 10 years, diabetes mellitus type 2, and chronically hemodialyzed for 7 years. The laboratory findings showed a serum creatinine kinase (CK) level of 473 IU/L, creatinine kinase-myocardial band (CK-MB) level of 6.4 ng/mL,

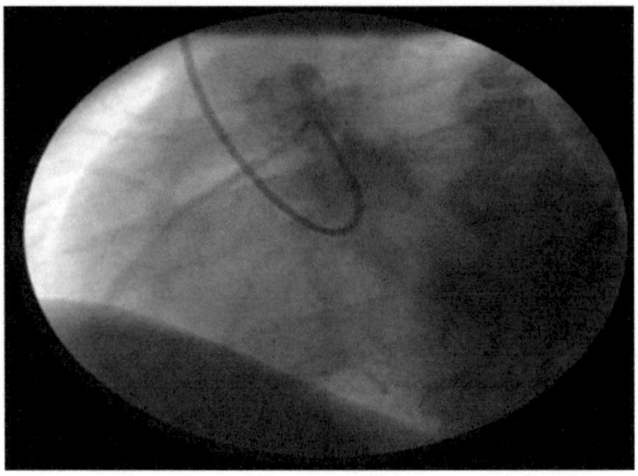

Figure 6.
Right coronary artery.

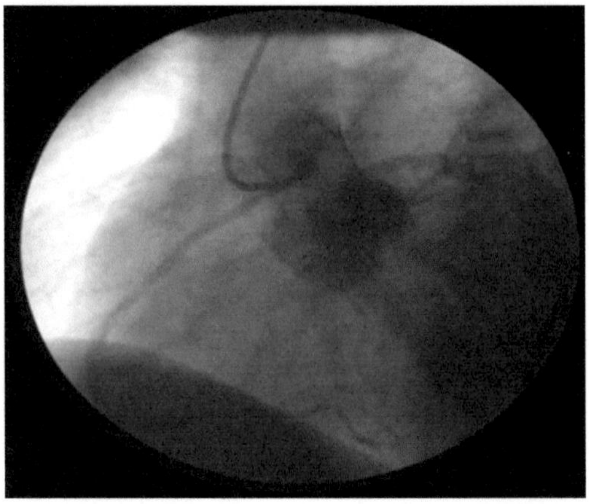

Figure 7.
Right coronary artery rising from left Valsalva sinus from a separate ostia with the left coronary artery.

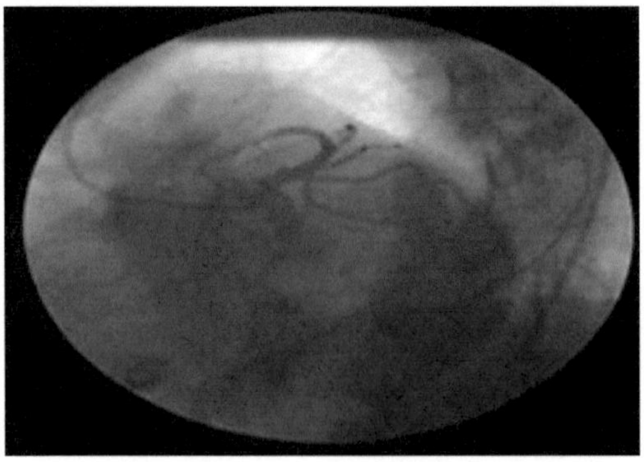

Figure 8.
Tortuous medial and distal segments of LAD.

and troponin-I level of 0.15 ng/mL. Electrocardiography demonstrated ST segment depression of 1–2 mm in leads V4–V6, and inverted T wave in D2, D3, aVF. The transthoracic echocardiography (TTE) revealed severely hypokinetic medioapical

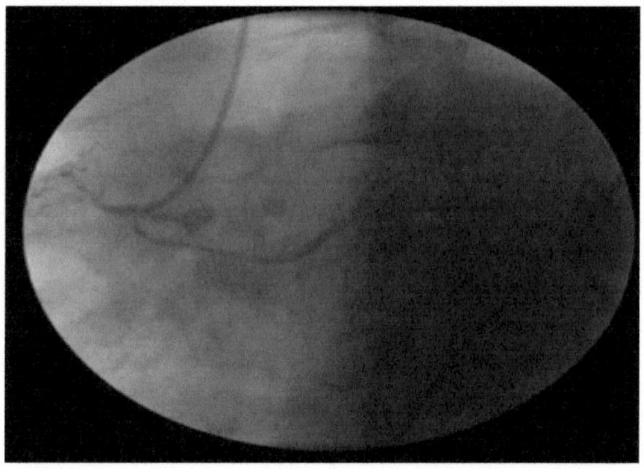

Figure 9.
Revealed LCx arising from right coronary sinus.

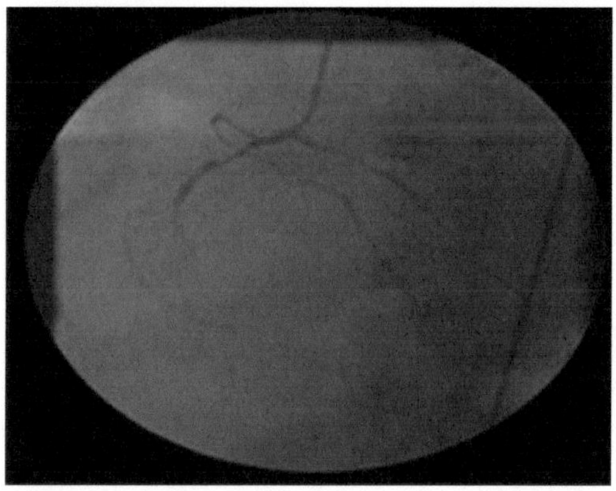

Figure 10.
Revealed a LCx as a proximal branch of RCA.

segments of anterolateral wall, and hypokinetic basal segments of interventricular septum and inferior wall.

The coronary angiography revealed a left circumflex artery (LCx) as a proximal branch of the right coronary artery (**Figure 10**). The LAD contained an proximal lesion up to 80%. The LCx and RCA are occluded in medial segment.

7. Comments

The coronary angiography of patients with coronary ischemia determined atherosclerotic disease with possibility of the presence of coronary artery anomalies, but also coronary angiography may reveal coronary artery anomaly without the presence of atherosclerotic changes. The ectopic origin from opposite sinus of coronary artery anomalies that presents with atherosclerotic changes continues to exist as a challenge during treatment in interventional cardiology.

The atherosclerotic coronary artery disease leads to the need of coronarography which can find out the presence of coronary artery anomalies. We should think about these anomalies during coronarography knowing that based on type of these anomalies and considering the vulnerability to atherosclerosis will be determined the method of the treatment.

Author details

Xhevdet Krasniqi* and Hajdin Çitaku
University Clinical Center of Kosova, Prishtina, Kosovo

*Address all correspondence to: xhevdeti_16@hotmail.com

References

[1] Pérez-Pomares JM, de la Pompa JL, Franco D, Henderson D, Ho SY, Houyel L, Kelly RG, Sedmera D, Sheppard M, Sperling S, Thiene G, van den Hoff M, Basso C. Congenital coronary artery anomalies: A bridge from embryology to anatomy and pathophysiology—A position statement of the development, anatomy, and pathology ESC Working Group. Cardiovascular Research. 2016;**109**(2):204-216

[2] Zhang LJ, Yang GF, Huang W, Zhou CS, Chen P, Lu GM. Incidence of anomalous origin of coronary artery in 1879 Chinese adults on dual-source CT angiography. Netherlands Heart Journal. 2010;**18**(10):466-470

[3] Yamanaka O, Hobbs RE. Coronary artery anomalies in 126,595 patients undergoing coronary arteriography. Catheterization and Cardiovascular Diagnosis. 1990;**21**(1):28-40

[4] Schwartz L, Kip KE, Alderman E, Lu J, Bates ER, Srinivas V, Bach RG, Mighton LD, Feit F, King S 3rd, Frye RL, BARI 2D Study Group. Baseline coronary angiographic findings in the Bypass Angioplasty Revascularization Investigation 2 Diabetes trial (BARI 2D). The American Journal of Cardiology. 2009;**103**(5):632-638

[5] Risau W, Flamme I. Vasculogenesis. Annual Review of Cell and Developmental Biology. 1995;**11**: 73-91

[6] Folkman J, Haudenschild C. Angiogenesis in vitro. Nature. 1980;**288**(5791):551-556

[7] Ogden J. The origin of coronary arteries. Circulation. 1998:38-150

[8] Carmeliet P. Mechanisms of angiogenesis and arteriogenesis. Nature Medicine. 2000;**6**:389-395

[9] Cai W, Schaper W. Mechanisms of arteriogenesis. Acta Biochimica et Biophysica Sinica. 2008;**40**:681-692

[10] Ouali S, Neffeti E, Sendid K, Elghoul K, Remedi F, Boughzela E. Congenital anomalous aoric origins of the coronary arteries in adults: A Tunisian coronary arteriography study. Archives of Cardiovascular Disease. 2009;**102**(3):201-208

[11] Angelini P. Coronary artery anomalies-current clinical issues: Definitions, classifications, incidence, clinical relevance, and treatment guidelines. Texas Heart Institute Journal. 2002;**29**:271

[12] Angelini P, Velasco JA, Flamm S. Coronary anomalies: Incidence, pathophysiology, and clinical relevance. Circulation. 2002;**105**(20):2449-2454

[13] Angelini P. Coronary artery anomalies: An entity in search of an identity. Circulation. 2007;**115**:1296-1305

[14] Kayalar N, Burkhart HM, Dearani JA, Cetta F, Schaff H. Congenital coronary anomalies and surgical treatment. Congenital Heart Disease. 2009;**4**:239-251

[15] Hofmeyr L, Moolman J, Brice E, Weich H. An unusual presentation of an anomalous left coronary artery arising from the pulmonary artery (ALCAPA) in adult: Anterior papillary muscle rupture causing severe mitral regurgitation. Echocardiography. 2009;**26**(4):474-477

[16] Yau JM, Singh R, Halpern EJ, Fischman D. Anomalous origin of the left coronary artery from the pulmonary artery in adults: A comprehensive review of 151 adult cases and a new diagnosis in a 53-year-old woman. Clinical Cardiology. 2011;**34**:2014-2010

[17] Basso C, Maron BJ, Corrado D, Thiene G. Clinical Profile of congenital coronary artery anomalies with origin from the wrong aortic sinus leading to sudden death in young competitive athletes. Journal of the American College of Cardiology. 2000;**35**:1493-1501

[18] Mangukia C. Coronary artery fistula. The Annals of Thoracic Surgery. 2012;**93**:2084-2092

[19] Cheng TO. Left coronary artery-to-left ventricular fistula. A follow-up report. International Journal of Cardiology. 2007;**118**(2):233

[20] Cebi N, Schulze-Waltrup N, Frömke J, Scheffold T, Heuer H. Congenital coronary artery fistulas in adults: Concomitant pathologies and treatment. The International Journal of Cardiovascular Imaging. 2008;**24**(4):349-355

[21] Alegria JR, Herrmann J, Holmes DR Jr, Lerman A, Rihal CS. Myocardial bridging. European Heart Journal. 2005;**26**(12):1159-1168

[22] Frommelt P, Frommelt M. Congenital coronary artery anomalies. Pediatric Clinics of North America. 2004;**51**:1273-1288

[23] Porta I, MacDonald ST, Selvanayagan JB, Banning AP. Intravascular ultrasound to guide stenting of an anomalous right coronary artery coursing between the aorta and pulmonary artery. The Journal of Invasive Cardiology. 2005;**17**(11):E33-E36

[24] Gambetta K, Cui W, el-Zein C, Roberson DA. Anomalous left coronary artery from the right sinus valsalva and noncompaction of the left ventricle. Pediatric Cardiology. 2008;**29**(2):434-437

[25] Luo L, Kebede S, Wu S, Stouffer GA. Coronary artery fistulae. The American Journal of the Medical Sciences. 2006;**332**(2):79-78

[26] Ishikawa Y, Akasaka Y, Suzuki K, Fujiwara M, Ogawa T, Ymazaki K, Niino H, Tanak M, Ogata K, Morinaga S, Ebihara Y, Kawahara Y, Sugiura H, Takimoto T, Komatsu A, Shinagawa T, Taki K, Satoh H, Yamada K, Yanagida-lida M, Shimokawa P, Shimada K, Nishimura C, Ito K, Ishii T. Anatomic properties of myocardial bridge predisposing to myocardial infarction. Circulation. 2009;**120**(5):376-383

[27] Pristipino C. Radial versus femoral access in ACSs. Lancet. 2011;**378**(9792):661

[28] Shah RM, Patel D, Abbate A, Cowley MJ, Jovin IS. Comparison of transradial coronary procedures via right versus left radial artery approach: A meta-analysis. Catheterization and Cardiovascular Interventions. 2016;**88**(7):1027-1033

[29] Hamood H, Makhoul N, Grenadir E, Kusniec F, Rosenschein U. Anchor wire technique improves device deliverability during PCI of CTOs and other complex subsets. Acute Cardiac Care. 2006;**8**(3):139-142

Chapter 8

Systemic Sclerosis

Murat Borlu and Eda Öksüm Solak

Abstract

Systemic sclerosis (SSc) is a chronic, autoimmune disease which can affect the blood vessels, the visceral organs, and the skin. SSc, most commonly, develops between the ages of 30 and 50, but it can be seen at any age. In terms of skin involvement, SSc can be classified as limited or diffuse. Its etiopathogenesis is still unclear. Microvascular dysfunction is thought to be followed by immunological activation, collagen and extracellular matrix deposition, and finally fibrosis. Diagnosis is based on clinical presentation. Sclerosis of the metacarpophalangeal and/or metatarsophalangeal joints is the major diagnostic criterion, whereas sclerodactylia, digital ulcers (DU), and pulmonary fibrosis are the minor criteria. SSc is diagnosed with one major criterion or two minor criteria. Detection of autoantibodies can help the diagnosis. Antinuclear antibody (ANA), anti-centromere antibody, anti-scl 70, RNA polymerase 1 and 3, and anti-fibrillin antibody can be found positive in SSc. SSc must be differentiated from all sclerosing diseases and the diseases with Raynaud's phenomenon. Visceral diseases, such as primary pulmonary hypertension, primary biliary cirrhosis, and infiltrative cardiomyopathy, should also be considered in its differential diagnosis. The main treatment goal is to target visceral involvement.

Keywords: sclerosis, chronic, microvascular, visceral

1. Introduction

Systemic sclerosis (SSc) is a chronic immune-mediated connective tissue disease, including the skin, inner organs, and blood vessels, with heterogeneous multiple organ involvement whose etiology is unknown. It has two subtypes: diffuse and limited. Its incidence ranges from 4 to 43 million people [1–3] per year, with a prevalence of 88–443 million [4, 5]. It can be seen at any age, though it is most commonly seen in patients aged 30–50 years. The disease shows an earlier onset and a more severe course in black patients [6]. The incidence in women is three to four times higher than in men [6]. Epidemiological studies showed that there is a significant increase in SSc risk in people, whose first-degree relatives have this disease [7].

2. Pathogenesis

The pathogenesis of SSc is not fully known. Disease-triggering agents are some organic solvents (e.g., silica, vinyl chloride, trichloroethylene, epoxy resins, benzene, carbon tetrachloride), some viral diseases (HSV5, CMV), some medications (bleomycin, pentazocine), and radiotherapy [8].

Basic pathogenicities of the disease are microvascular function disorder (vasculopathy) and immune activation, and the final effect of these events is progressive tissue fibrosis with activation of fibroblasts [9].

Vascular disease is observed earliest in SSc pathogenesis. The disease plays an important role in the occurrence of pulmonary artery hypertension, Raynaud's phenomenon, renal damage, and digital ulcers (DU) [10–13]. Raynaud's phenomenon, a typical clinical characteristic of SSc, is a finding characterized by persistent vasospasm and increased adhesion molecules following ischemia and reperfusion attacks [14]. Increased adhesion molecules trigger platelets and neutrophils that bind to endothelial cells and produce superoxide radicals responsible for endothelial cell damage [15]. In addition, an imbalance between vasoconstrictor agents such as endothelin-1 (ET-1) and vasodilating agents such as nitric oxide was observed in SSc, which plays a role in the change of vascular permeability [16]. Increased ET-1 expression plays a role in vascular fibrosis, inflammation, and increased smooth muscle cell proliferation [12]. An impaired cross talk between endothelial cells and perivascular cells may induce an abnormal expression of endothelial growth factor (VEGF), TGF-β, and platelet-derived growth factor (PDGF) in SSc. This may lead to a disruption of peripheral vascularization, which results in fibrosis of the skin and internal organs [17].

It has been determined that Th2 cytokines such as IL 4, IL 5, and IL 13 are over-secreted in SSc patients. It increases IL 4 collagen synthesis and TGF-β production. With IL 13, however, fibroblast activity and TGF-β stimulation are carried out [18]. Also, B-cells produce antibodies and carry out direct fibroblast stimulation via IL 6 [18].

3. Diagnostic criteria and classification

SSc is a heterogeneous disease and this disease spectrum contains different forms. Although the CREST syndrome (calcinosis, Raynaud's phenomenon, esophageal motility disorder, sclerodactyly, telangiectasia) was first defined in 1959,

Criteria	Score
1. Skin hardening of the fingers of both hands spreading proximal to the MKF joints	9
2. Skin hardening of the fingers	
• Puffy fingers	2
• Sclerodactyly (only high scores will be counted)	4
3. Fingertip lesions	
• Digital ulcers	2
• Digital pitting (only high scores will be counted)	3
4. Telangiectasia	2
5. Capillary changes such as abnormal fingernails	2
6. Pulmonary arterial hypertension	2
• Interstitial lung disease (maximum 2 points)	2
7. Raynaud's phenomenon	3
8. SSc-related autoantibodies (anti-centromere, anti-topoisomerase I, anti-RNA polymerase III)	3

Table 1.
ACR/EULAR 2013 systemic sclerosis classification criteria (adapted from source [21]).

it does not fit any classification criteria, and since a subgroup has not been fully defined, it is recommended not to use it nowadays [19].

The criteria published by the American Radiology Associates (ARA) in 1980 have been used for classification for a long time. According to this, symmetrical skin sclerosis in the metacarpophalangeal joint or proximal of the metatarsophalangeal joint is a major criterion, while sclerodactylia, digital atrophic cicatrice, and the loss of finger fat tissue in the distal and bilateral fibrosis in the lungs constitute minor criteria. Diagnosis is possible with the presence of the major criteria or two minor criteria [20]. The ARA criteria do not include patients with limited skin involvement or early SSc and do not involve capillaroscopy/autoantibody tests.

In 2013, the American College of Rheumatology/European League against Rheumatism (ACR/EULAR) classification criteria were defined. Sensitivity and specificity of these criteria were shown to be higher than the ARA criteria [21]. **Table 1** shows the ACR/EULAR 2013 systemic sclerosis classification criteria. The total score is the sum of the points received from different categories. Patients with a score of nine or higher are classified as definitive systemic sclerosis. However, the use is not recommended in patients with skin thickening that does not involve the fingers, in the presence of scleroderma-like diseases or better clinical pictures.

4. Clinical findings

4.1 Cutaneous involvement

Systemic sclerosis is divided into two subgroups, according to cutaneous involvement: limited systemic sclerosis and diffuse systemic sclerosis. If cutaneous involvement is limited to distal extremities and the face, it is classified as limited systemic sclerosis, and if the involvement is present in the truncus and extremities, it is classified as diffuse systemic sclerosis [22]. Also, systemic sclerosis sine scleroderma, of which 5% of SSc patients are affected, shows typical SSc symptoms but no fibrosis of the skin [23]. Raynaud's phenomenon is seen in more than 90% of SSc patients and is triggered by exposure to cold or emotional stress. It is characterized by white, blue, and red discoloration after triggering and most often affects the hands, feet, tongue, ears, and nose [24].

Cutaneous involvement generally consists of three phases. In the first stage, the edematous phase, non-pitting edema, facial mask appearance, and swelling of the fingers can be seen. After that comes the indurative phase, where the skin hardens and gets a shiny and tense appearance. The last stage is the atrophic phase, characterized by claw hand and sclerodactyly. Sharpening of the nose, thinning of the lips, vertical streaks around the mouth (**Figure 1**), and facial mimic loss may be seen. Skin lines or hyperpigmentation of skin areas exposed to trauma and depigmentation areas on the truncus, face, hand back, and leg fronts may develop. In some cases, the presence of depigmented areas, where perifollicular areas are preserved, results in an appearance of "salt and pepper" on the skin.

In systemic sclerosis disease-specific capillary dilatation, stumps and the presence of avascularity areas can be shown with nail bed capillaroscopy [25]. Sweat and atrophy of the sebaceous gland may lead to dry skin, flaking, and itching. Depending on the calcium accumulation in the skin, hard subcutaneous nodules may appear around the small joints of the hand (calcinosis cutis), which may open to the outside and become ulcers [26]. Ulcers (**Figure 2**) may also develop due to ischemia, trauma, and fibrotic tissue. Telangiectasia is commonly seen in SSc and is stated in the classification criteria [21].

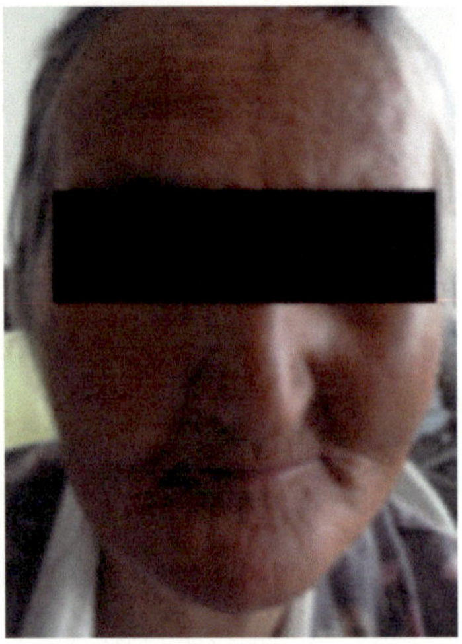

Figure 1.
Nasal sharpening and perioral vertical streaks in a patient.

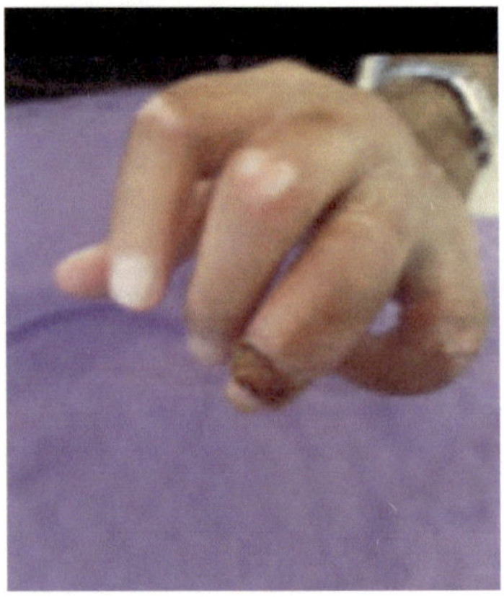

Figure 2.
Digital ulcer in a patient.

4.2 Gastrointestinal system involvement

Ninety percent of scleroderma patients show gastrointestinal involvement. Although it can involve any part of the gastrointestinal tract from the mouth to the anus, the esophagus is the most commonly involved area [27]. The most common complications of SSc in the oral cavity are microstomy and xerostomy. Symptoms of esophageal disease depend on dysmotility and reflux, and related to this, dysphagia, odynophagia, regurgitation, pyrosis, chronic cough, or hoarseness can occur. Stricture, Barrett's esophagus, aspiration pneumonia, and adenocarcinoma may develop as complications in these patients [27, 28]. The two most common symptoms of SSc in the stomach are gastroparesis and gastric antral vascular ectasia (GAVE). A typical striped watermelon appearance is present. Iron deficiency anemia due to GAVE may also be seen [28]. After bacterial growth due to small intestinal involvement, diarrhea and malabsorption or pseudo-obstruction due to hypomotility and dilatation may be observed [28]. Constipation, fecal, rectal prolapse, spontaneous perforation, and colon infarction may develop in the colon and have anorectal involvement [29].

4.3 Pulmonary involvement

Pulmonary involvement, the most important cause of mortality and morbidity in SSc, may occur in the form of interstitial lung disease (ILD) or pulmonary arterial hypertension (PAH) [30].

The highest risk for ILD development is within the first 5 years of the disease. Progressive exercise dyspnea and nonproductive cough are the most common symptoms. Diagnosis is placed with the help of imaging methods and pulmonary function tests. High-resolution computed tomography is a sensitive method in imaging and can show the degree of fibrosis.

Pulmonary artery hypertension affects approximately 15% of scleroderma patients and leads to right heart failure, and the gold standard method for the diagnosis of PAH is right heart catheterization with high pressure in the pulmonary artery [31].

4.4 Musculoskeletal involvement

The most common involvement of the musculoskeletal system in SSc are tendinopathy, joint contractures, and in some cases arthritis. In 45–90% of patients, arthralgia and arthritis of the small joints of the wrist, knees, and ankles occur [31].

Tenosynovitis and tendon ruptures are frequently detected in SSc. Tendon friction sound that can be detected by physical examination is caused by fibrotic deposits in the tendon sheath [32].

Calcinosis and acroosteolysis in the bones (resorption of terminal phalanx) can be seen in SSc patients. These changes are related to digital ischemia [31].

In addition, muscle weakness and pain, which mainly affects the proximal muscles, is another common symptom in SSc. Muscle involvement may occur in the form of myopathy or myositis, and patients should be evaluated from this angle [31]. Muscle weakness may result from nonuse, sedentary life, or joint/tendon involvement or sometimes as a side effect of treatment.

Although respiratory muscle involvement is not common in SSc, respiratory muscles may be affected in SSc-polymyositis/dermatomyositis overlap syndrome [31].

4.5 Cardiac involvement

Cardiac involvement in SSc can be seen in two ways: primary and secondary. Secondary involvement may be secondary to pulmonary arterial hypertension, interstitial lung disease, or kidney disease [33]. Primarily, all layers of the heart, conduction system, coronary vessels, and valves can be involved. Because of these involvements, pericardial effusion, supraventricular or ventricular arrhythmias, conduction disorders, valve regurgitation, myocardial ischemia, myocardial hypertrophy, and heart failure may develop [34].

Because myocardial findings are often faint, especially in the early stage, it is very difficult to detect these patients. However, when the involvement progresses to the symptomatic stage, the patient develops one of the poor prognostic factors [31]. Therefore, early diagnosis of cardiac involvement in SSc is very important. Electrocardiography or echocardiography is not effective in detecting cardiac fibrosis. Magnetic resonance imaging (MRI) is the only method that can detect cardiac involvement in the early stages of the disease. A cardiac MRI can show myocardial inflammation, fibrosis, decreased perfusion of the heart muscle, and ventricular dysfunction [35].

4.6 Renal involvement

Renal involvement in scleroderma is quite common. Even though it appears often as mild renal dysfunction, it can also cause a severe clinical table called scleroderma renal crisis (SRC). The pathogenesis of SRC is not fully known, but studies suggest vasculopathy as a source. Corticosteroids and vasospasm-causing drugs (tacrolimus, cyclosporine, and cocaine) may play a role in the etiology [36]. The risk of SRC development increases in the presence of the autoantibodies anti-RNA polymerase III, anti-topoisomerase I, and anti-U3RNP [36].

Some patients show a chronic clinical table with gradual decrease in eGFR, increase in serum creatinine concentration, proteinuria, hematuria, and moderate arterial hypertension [31].

Decreased glomerular filtration rate, increased serum creatinine, hemolytic anemia, proteinuria, and decreased platelet count are laboratory findings indicating renal involvement [31].

5. Autoantibodies

Antinuclear antibodies (ANA) are 90–95% positive in SSc patients and are the most commonly detected autoantibody. Scleroderma-like diseases should also be considered in the case of ANA negativity. Anti-topoisomerase I (anti-scl 70) antibodies are connected to pulmonary complications, digitals ulcers, and progressive hand involvement. Anti-centromere antibodies are common in limited SSc and positively increase the risk of pulmonary fibrosis and pulmonary hypertension. Anti-RNA polymerase III antibody is associated with renal crisis. Also anti-U3RNP and anti-Th/To antibodies can be detected positively, and the anti-U3RNP antibody increases the risk of pulmonary artery hypertension and cardiovascular complications [37].

6. Histopathology

The diagnosis of systemic sclerosis is placed clinically; therefore biopsy is not recommended routinely [38]. It can be used to rule out other diseases for differential diagnosis. Histologically, excessive collagen accumulation, atrophy of pilosebaceous and eccrine glands, subcutaneous fat loss, and lymphocytic infiltrate are observed. Increased collagen can compress adnexal structures, especially eccrine glands [38].

7. Differential diagnosis

There are many diseases that may trigger dermal sclerosis. The form and character of skin involvement, history of underlying diseases, and chemical exposures are helpful factors in approaching the diagnosis of a patient with skin thickening. Some laboratory studies may be beneficial in verifying imaging and skin biopsy diagnosis [8]. Eosinophilic fasciitis, scleroderma, scleromyxedema, and nephrogenic systemic fibrosis are important in the differential diagnosis of SSc. SSc can be differentiated from other scleroses by the presence of Raynaud's phenomenon, typical distal extremity involvement, nail fold capillary findings, presence of autoantibodies, and internal organ involvement. While the groove mark, which is a recess caused by the withdrawal of subcutaneous tissues along the path of the superficial vessels, favors the eosinophilic fasciitis [39], the detection of monoclonal gammopathy directs the diagnosis towards scleroderma or scleromyxedema (bbb). Again, if underlying renal failure or gadolinium exposure is detected, the first thought should be towards nephrogenic systemic fibrosis [40].

8. Treatment

The pathogenesis of SSc is still unclear. In recent years, advances have been made in the treatment of the disease, resulting in a prominent improvement in survival rates. The efficient treatment of complications increases the chances of success. Disease duration, complications, and disease activity should be taken into consideration when making therapeutic decisions. The treatment is based on modifying agents and organ-specific drugs [41].

Peripheral vascular involvement frequently occurs as Raynaud's phenomenon. In addition, digital ischemia due to digital vasculopathy, digital ulcers, and associated amputations may be the causes for morbidity in SSc.

Patients with Raynaud's phenomenon should protect themselves from the cold. They also should not smoke and should avoid vasoconstrictor agents. Calcium channel inhibitors (nifedipine, diltiazem, amlodipine) should be the first choice as treatment [41]. Iloprost and other intravenous prostanoids can be used in cases that do not respond to the above. Also, phosphodiesterase type 5 (PDE5) inhibitors may be effective in resistant cases. Selective serotonin reuptake inhibitors, pentoxifylline, prazosin, and endothelin receptor antagonists (ERA) are agents that are less effective and used in selected cases [42].

Early treatment of patients with digital ischemia SSc reduces the risk of morbidity. Intermittent infusion of prostacyclin or analogs was found to be effective in the treatment of RF and ischemic digital ulcer [41]. Sildenafil and bosentan are recommended for the treatment of digital ulcers that developed due to unsuccessful treated systemic sclerosis with calcium channel blockers and prostanoid therapies [41]. There are also uncontrolled studies suggesting that atorvastatin, vitamin E, and intravenous N-acetylcysteine may be beneficial. Mesenchymal stem cell therapy is one of the developing methods in the treatment of DU [43].

The first choice for the treatment of pulmonary artery hypertension is phosphodiesterase type 5 inhibitors (e.g., sildenafil or tadalafil) or endothelin receptor antagonists such as bosentan and macitentan. If those are not effective, prostanoids may be added to the treatment [44]. Prostanoids are the first choice in severe cases. If there is no response, combination treatments can be applied. Recently, the guanylate cyclase agonist riociguat has been involved in treatment [41].

The treatment of interstitial lung disease patients is based on immunosuppressive drugs. The first preferred agent is oral mycophenolate mofetil (MMF). One of the applied treatment regimens for nonresponsive patients is the administration of cyclophosphamide (CYC) orally at doses of 1–2 mg/kg/day or iv 600 mg/m^2/month. After the disease activity has been taken under control, it is recommended to continue treatment with azathioprine at a dose of 2.5 mg/kg/day [41]. In selected patients, rituximab (RTX, anti-CD20 monoclonal antibody) can be used as an alternative treatment [41, 45].

Immunosuppressants such as methotrexate (MTX), cyclophosphamide, and mycophenolate mofetil are commonly used in fibrosis of the skin [31]. If the treatment is unsuccessful or if these drugs cannot be used for whatever reason, low-dose systemic corticosteroids or rituximab may be preferred [31]. The effect of D-penicillamine, which has been used for many years, is controversial [46].

In systemic sclerosis, exertional dyspnea, tachycardia, and chest pain may occur due to myocardial involvement. Selective beta-blockers are effective in this kind

Organ involvement	Recommendation	Strength of recommendation
Raynaud's phenomenon	• Calcium antagonists	A
	• PDE-5 inhibitors	A
	• Prostanoids	A
	• Fluoxetine	C
Digital ulcers	• Intravenous iloprost	A
	• PDE-5 inhibitors	A
	• Bosentan*	A
Pulmonary arterial hypertension	• ERA, PDE-5 inhibitors, riociguat	B
	• Intravenous epoprostenol	A
	• Other prostacyclin analogs (iloprost, treprostinil)	B
Skin and lung disease		
Skin	• Methotrexate	A
Lung disease	• Cyclophosphamide	A
	• HSCT	A
SRC	• ACE inhibitors	**C**
	• **Glucocorticoids are associated with a higher risk of SRC	**C**
Gastrointestinal disease	PPI	B
	Prokinetic drugs	C
	Antibiotics	L

ERA, endothelin receptor antagonists; HSCT, hematopoietic stem cell transplantation; PAH, pulmonary arterial hypertension; PDE-5, phosphodiesterase type 5; PPI, proton pump inhibitor; SRC, scleroderma renal crisis; SSc, systemic sclerosis; RP, Raynaud's phenomenon.
Treats and reduces the formation of new ulcers.
**Blood pressure and renal function should be carefully monitored in SSc patients treated with glucocorticoids.*

Table 2.
The updated EULAR recommendations for treatment of systemic sclerosis.

of symptoms [33]. If cardiac tamponade develops due to pericarditis in cardiac involvement, treatment becomes more difficult. The patient may not respond to systemic corticosteroid therapy, and drainage treatment may become necessary [47]. Some SSc patients may develop microvascular ischemia due to vasospasm of small coronary arteries and arterioles, also called cardiac Raynaud's phenomenon. Nifedipine treatment is quite effective in such patients [48].

SSc patients with GI involvement of dysphagia, pyrosis, esophageal reflux, esophagitis, distention, abdominal pain, and diarrhea can be treated with proton pump inhibitors (PPI), prokinetic drugs, and intermittent antibiotics (gg). Patients that developed gastric antral vascular ectasia may have severe upper GI bleeding. Here, supportive treatment and endoscopic treatment methods can be used. Surgical antrectomy becomes necessary as the last resort in resistant cases [49].

The first choice for patients with scleroderma renal crisis is a high dose of angiotensin receptor antagonists (ACE-I). It was determined that this treatment significantly decreased mortality [31]. In cases with insufficient response, angiotensin receptor blockers (ARB) and calcium channel blockers may be combined with ACE-I treatment of nitrates [31]. Beta-blockers are not recommended due to

DOI: http://dx.doi.org/10.5772/intechopen.91318

their vasoconstriction enhancing effects. Hypotension should be avoided, and close monitoring should be performed for patients using systemic steroids since they significantly increase the risk [50].

The updated EULAR recommendations for the treatment of systemic sclerosis are summarized in **Table 2**.

9. Result

As a result, we should not forget that systemic sclerosis is a chronic disease that can cause serious morbidity and mortality and that placing a diagnosis can be difficult from time to time. The disease does not have a specific treatment. The patient should be evaluated according to involved systems and clinical table. In addition, early diagnosis and early initiation of treatment increase the chances of efficacy. A better understanding of the pathogenesis may lead to the development of more successful and specific treatment methods.

Author details

Murat Borlu* and Eda Öksüm Solak
Dermatology and Venereology Department, Faculty of Medicine, Erciyes University, Kayseri, Turkey

*Address all correspondence to: murat_borlu@hotmail.com

References

[1] Silman AJ, Jannini S, Symmons DPM, et al. An epidemiological study of scleroderma in the West Midlands. British Journal of Rheumatology. 1988;**27**:286-290. DOI: 10.1093/romatoloji/27.4.286

[2] Kaipiainen-Seppanen O, Aho K. Incidence of rare systemic rheumatic and connective tissue diseases in Finland. Journal of Internal Medicine. 1996;**240**:81-84. DOI: 10.1046/j.1365-2796.1996.14843000.x

[3] Lo Monaco A, Bruschi M, La Corte R, et al. Epidemiology of systemic sclerosis in a district of northern Italy. Clinical and Experimental Rheumatology. 2011;**29**(Suppl 65):S10-S14

[4] Bernatsky S, Joseph L, Pineau CA, et al. Scleroderma prevalence: Demographic variations in a population-based sample. Arthritis and Rheumatism. 2009;**61**:400-404. DOI: 10.1002/art.24339

[5] Allcock RJ, Forrest I, Corris PA, et al. A study of the prevalence of systemic sclerosis in Northeast England. Rheumatology (Oxford). 2004;**43**:596-602. DOI: 10.1093/rheumatology/keh124

[6] Mayes MD, Lacey JV, Beebe-dimmer J, et al. Prevalence, incidence, survival, and disease characteristics of systemic sclerosis in a large US population. Arthritis and Rheumatism. 2003;**48**:2246-2255. DOI: 10.1002/art.11073

[7] Agarwal SK, Tan FK, Arnett FC. Genetics and genomic studies in scleroderma (systemic sclerosis). Rheumatic Diseases Clinics of North America. 2008;**34**:17-40. DOI: 10.1016/j.rdc.2007.10.001

[8] Jimenez SA, Koenig AS. Scleroderma. Emedicine. 2007. Available from: http://www.emedicine.com/med/topic2076.htm

[9] Denton CP, Black CM, Korn JH. Systemic sclerosis: Current pathogenetic concepts and future prospects for targeted therapy. Lancet. 1996;**347**:1453-1458. DOI: 10.1016/S0140-6736(96)91687-6

[10] Matucci-Cerinic M, Kahaleh B, Wigley FM. Review: Evidence that systemic sclerosis is a vascular disease. Arthritis and Rheumatology. 2013;**65**:1953-1962. DOI: 10.1002/art.37988

[11] Kahaleh MB. Vascular involvement in systemic sclerosis (SSc). Clinical and Experimental Rheumatology. 2004;**22**(Suppl 33):S19-S23

[12] Abraham D, Distler O. How does endothelial cell injury start? The role of endothelin in systemic sclerosis. Arthritis Research & Therapy. 2007;**9**(Suppl 2):S2. DOI: 10.1186/ar2186

[13] Cantatore FP, Maruotti N, Corrado A, et al. Angiogenesis dysregulation in the pathogenesis of systemic sclerosis. BioMed Research International. 2017;**2017**: Article ID 5345673, 6 pages. DOI: 10.1155/2017/5345673

[14] Rabquer BJ, Koch AE. Angio genesis and vasculopathy in systemic sclerosis: Evolving concepts. Current RheumatologyReports. 2012;**14**(1):56-63. DOI: 10.1007/s11926-011-0219-1

[15] Pattanaik D, Brown M, Postlethwaite BC, et al. Pathogenesis of systemic sclerosis. Frontiers in Immunology. 2015;**6**: article no. 272. DOI: 10.3389/fimmu.2015.00272

[16] Kahaleh B. Vascular disease in scleroderma: Mechanisms of vascular injury. Rheumatic Disease Clinics of

North America. 2008;**34**(1):57-71. DOI: 10.1016/j.rdc.2007.12.004

[17] Cipriani P, DiBenedetto P, Ruscitti P, et al. Impaired endothelium-mesenchymal stem cells cross-talk in systemic sclerosis: A link between vascular and fibrotic features. Arthritis Research & Therapy. 2014;**16**:442. DOI: 10.1186/s13075-014-0442-z

[18] Varga J, Abraham D. Systemic sclerosis: A prototypic multisystem fibrotic disorder. Journal of Clinical Investigation. 2007;**117**:557-567. DOI: 10.1172/JCI31139

[19] Wollheim FA. Classification of systemic sclerosis. Visions and reality. Rheumatology. 2005;**44**(10):1212-1216. DOI: 10.1093/rheumatology/keh671

[20] American Rheumatism Association. Preliminary criteria for the classification of systemic sclerosis (scleroderma). Subcommittee for the scleroderma criteria of the American rheumatism association diagnostic and therapeutic criteria committee. Arthritis and Rheumatism. 1980;**23**(5):581-590. DOI: 10.1002/art.1780230510

[21] Van den Hoogen F, Khanna D, Fransen J, et al. 2013 classification criteria for systemic sclerosis: An American College of Rheumatology/European League Against Rheumatism collaborative initiative. Annals of the Rheumatic Diseases. 2013;**72**(11):1747-1755. DOI: 10.1002/art.38098

[22] Barsotti S, Bellando Randone S, Guiducci S, et al. Systemic sclerosis: A critical digest of the recent literature. Clinical and Experimental Rheumatology. 2014;**32**(Suppl 86): S194-S205

[23] Kucharz EJ, Kopeć-Mędrek M. Systemic sclerosis sine scleroderma. Advances in Clinical and Experimental Medicine. 2017;**26**:875-880. DOI: 10.17219/acem/64334

[24] Herrick AL. The pathogenesis, diagnosis and treatment of Raynaud phenomenon. Nature Reviews Rheumatology. 2012;**8**:469-479. DOI: 10.1038/nrrheum.2012.96

[25] Cutolo M, Pizzorni C, Tuccio M, et al. Nail fold video capillaroscopic patterns and serum auto antibodies in systemic sclerosis. Rheumatology (Oxford). 2004;**43**:719-726. DOI: 10.1093/rheumatology/keh156

[26] Ferreli C, Gasparini G, Parodi A, et al. Cutaneous manifestations of scleroderma and scleroderma-like disorders: A comprehensive review. Clinical Reviews in Allergy and Immunology. 2017;**53**:306-336. DOI: 10.1007/s12016-017-8625-4

[27] Shreiner A, Murray C, Denton C, et al. Gastrointestinal manifestations of systemic sclerosis. Journal of Scleroderma and Related Disorders. 2016;**1**:247-256. DOI: 10.5301/jsrd.5000214

[28] McFarlane IM, Bhamra MS, Kreps A, et al. Gastrointestinal manifestations of systemic sclerosis. Rheumatology (Sunnyvale). 2018;**8**(1):235. DOI: 10.4172/2161-1149.1000235

[29] Emmanuel A. Current management of the gastrointestinal complications of systemic sclerosis. Nature Reviews. Gastroenterology & Hepatology. 2016;**13**:461-472. DOI: 10.1038/nrgastro.2016.99

[30] Solomon JJ, Olson AL, Fischer A, et al. Scleroderma lung disease. European Respiratory Review. 2013;**22**:6-19. DOI: 10.1183/09059180.00005512

[31] Sobolewski P, Maślińska M, Wieczorek M, et al. Systemic sclerosis—Multidisciplinary disease: Clinical features and treatment. Reumatologia. 2019;**57**(4):221-233. DOI: 10.5114/reum.2019.87619

[32] Clements PJ, Allanore Y, Khanna D, et al. Arthritis in systemic sclerosis: Systematic review of the literature and suggestions for the performance of future clinical trials in systemic sclerosis arthritis. Seminars in Arthritis and Rheumatism. 2012;**41**(6):801-814. DOI: 10.1016/j.semarthrit.2011.10.003

[33] Kahan A, Coghlan G, McLaughlin V. Cardiac complications of systemic sclerosis. Rheumatology (Oxford). 2009;**48**(3):45-48. DOI: 10.1093/rheumatology/kep110

[34] Desai CS, Lee DC, Shah SJ. Systemic sclerosis and the heart: Current diagnosis and management. Current Opinion in Rheumatology. 2011;**23**:545-554. DOI: 10.1097/BOR.0b013e32834b8975

[35] Fernández-Codina A, Simeón-Aznar CP, Pinal-Fernandez I, et al. Cardiac involvement in systemic sclerosis: Differences between clinical subsets and influence on survival. Rheumatology International. 2017;**37**:75-84. DOI: 10.1007/s00296-015-3382-2

[36] Bose N, Chiesa-Vottero A, Chatterjee S. Scleroderma renal crisis. Seminars in Arthritis and Rheumatism. 2015;**44**:687-694. DOI: 10.1016/j.semarthrit.2014.12.001

[37] Castro SV, Jimenez SA. Biomarkers in systemic sclerosis. Biomarkers in Medicine. 2010;**4**:133-147. DOI: 10.2217/bmm.09.79

[38] Laga AC, Larson S, Granter SR. Histopathologic spectrum of connective tissue disease commonly affecting the skin. Surgical Pathology Clinics. 2017;**10**:477-503. DOI: 10.1016/j.path.2017.01.012

[39] Bischoff L, Derk CT. Eosinophilic fasciitis: Demographics, disease pattern and response to treatment: Report of 12 cases and review of the literature. International Journal of Dermatology. 2008;**47**(1):29-35. DOI: 10.1111/j.1365-4632.2007.03544.x

[40] Nashel J, Steen V. Scleroderma mimics. Current Rheumatology Reports. 2012;**14**:39-46. DOI: 10.1007/s11926-011-0220-8

[41] Kowal-Bielecka O, Fransen J, Avouac J, et al. EUSTAR coauthors update of EULAR recommendations for the treatment of systemic sclerosis. Annals of the Rheumatic Diseases. 2017;**76**:1327-1339. DOI: 10.1136/annrheumdis-2016-209909

[42] Marvi U, Chung L. Digital ischemic loss in systemic sclerosis. International Journal of Rheumatology. 2010;**2010**: pii 130717. DOI: 10.1155/2010/130717

[43] Tyndall A. Successes and failures of stem cell transplantation in auto immune diseases. Hematology. American Society of Hematology. Education Program. 2011;**2011**:280-284. DOI: 10.1182/asheducation-2011.1.280

[44] Galiè N, Hoeper MM, Humbert M, et al. Guidelines for the diagnosis and treatment of pulmonary hypertension. European Respiratory Journal. 2009;**34**(6):1219-1263. DOI: 10.1183/09031936.00139009

[45] Antic M, Distler JH, Distler O. Treating skin and lung fibrosis in systemic sclerosis: A future filled with promise? Current Opinion in Pharmacology. 2013;**13**(3):455-462. DOI: 10.1016/j.coph.2013.05.016

[46] Furst DE, Clements PJ. D-penicillamine is not an effective treatment in systemic sclerosis. Scandinavian Journal of Rheumatology. 2001;**30**(4):189-191. DOI: 10.1080/030097401316909495

[47] Byers RJ, Marshall DA, Freemont AJ. Pericardial involvement in systemic sclerosis. Annals of the Rheumatic

Diseases. 1997;**56**(6):393-394. DOI: 10.1136/ard.56.6.393

[48] Lekakis J, Mavrikakis M, Emmanuel M, et al. Cold-induced coronary Raynaud's phenomenon in patients with systemic sclerosis. Clinical and Experimental Rheumatology. 1998;**16**(2):135-140

[49] Ingraham KM, O'Brien MS, Shenin M, Derk CT, Steen VD. Gastric antral vascular ectasia in systemic sclerosis: Demographics and disease predictors. Journal of Rheumatology. 2010;**37**(3):603-607. DOI: 10.3899/jrheum.090600

[50] Shanmugam VK, Steen VD. Renal disease in scleroderma: An update on evaluation, risk stratification, pathogenesis and management. Current Opinion in Rheumatology. 2012;**24**(6):669-676. DOI: 10.1097/BOR.0b013e3283588dcf

Chapter 9

Vasculitis and Vasculopathies

Ana Paula Urzedo, Mariane Martins Manso and Thaisa Macarini de Faria

Abstract

Our organism, as complex as it is, needs a giant vascular network to deliver nutrients to all cells, so vasculopathies and vasculitis are diseases present in all medical specialties. The skin and subcutaneous cellular tissue are irrigated by a vast vascular network, with cutaneous involvement related to these frequent pathologies. These can be restricted to the integumentary system or be part of systemic diseases with cutaneous manifestations, which make them of great interest to dermatologists. They can affect any caliber of vessels and present with several dermatological manifestations such as erythema, livedo reticularis, palpable purpura, nodules, ulcers, urticaria, hemorrhagic blisters, gangrene and other manifestations that can be isolated or associated with systemic signs and symptoms. However, there is no worldwide consensus regarding the classification of vasculitis, and the classification proposed in this chapter is based on the International Chapel Hill Conference Nomenclature of Vasculitides 2012, which is based on the size of the vessels. The purpose of this chapter is to compile a review of the most current treatments for these conditions.

Keywords: vasculitis, lymphocytic, cutaneous small vessel, systemic vasculitis, leukocytoclastic, cutaneous, vascular diseases

1. Introduction

Vasculitis and vasculopathies are a group of diseases that course with an inflammatory process, which attacks vascular endothelium of vessels of different calibers. Usually, they course with dermatology lesions, the diagnosis is difficult, and they have a few treatment options. Despite that, an early diagnosis can contribute to a better quality of patient's life and prevent further complications (**Figure 1**) [1].

2. Methods

The authors performed a bibliographic review from cutaneous vasculitis and vasculopathies from literature, including online academics platforms such as PubMed and Google Scholar and dermatology books as Rivitti and Fitispack. The keywords used from this chapter were cutaneous vasculitis, Takayasu arteritis, Giant Cell Arteritis, Polyarteritis Nodosa, Kawasaki Disease, Microscopic polyangiitis, Granulomatosis with Polyangiitis (Wegener's), Eosinophilic Granulomatosis with Polyangiitis (Churg-Strauss), Cryoglobulinemic Vasculitis, Urticaria Vasculitis, Henoch-Schönlein, Behçet's syndrome, Cutaneous leukocytoclastic angiitis, cutaneous vasculopathies,

IntechOpen

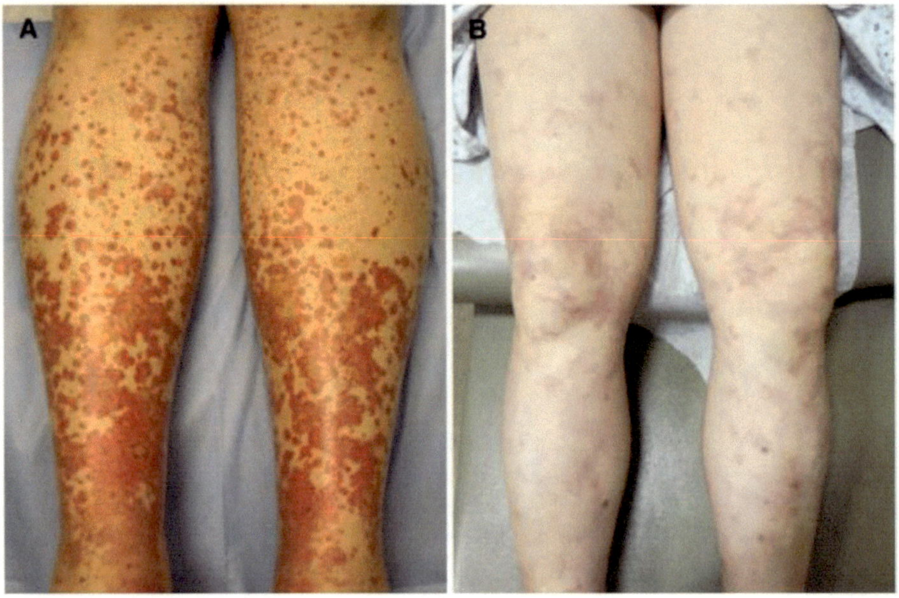

Figure 1.
Purpuric macules and papules coalescing into patches (A) in small vessel vasculitis and fixed livedo reticularis and subcutaneous nodules (B) in medium vessel vasculitis (polyarteritis nodosa) [2].

Phospholipid Antibody Syndrome, Blue Finger Syndrome, Acrocyanosis and others. The final objective was summing up the dermatology vasculitis and vasculopathies and helping physicians in early diagnosis and treatment.

3. Cutaneous vasculitis

Cutaneous vasculitis may affect vessels of different calibers, especially small and medium in the skin and the subcutaneous tissue, resulting in just a small cutaneous lesion until serious systemic commitment. Nevertheless, the classification is not a consensus. In general, they are classified using clinical criteria, size of the vessels, histopathological exams, laboratorial findings and etiologic agents [3, 4]. In 2012, the *Chapel Hill Consensus Conference (CHCC)* proposed a new classification from the most used since 1994, removing many eponymous. The classification is a strategic pillar to help choose the right path to lead these patients through the right treatment [4, 5] (**Table 1**).

3.1 Large vessel vasculitis

3.1.1 Takayasu arteritis

It is a large vessel vasculitis whose etiologic agent is unknown. It is more frequently found in young Asiatic women, in the proportion from 5 to 12 W:1 M, and usually affects the aorta and its main branches. It is not a rare disease but is not as usual as arteritis of giant cells. The diagnosis is usually late because these patients develop many collateral arteries, they have nonspecific symptoms and the development of these affections is very slow. Pregnancy in patients affected by Takayasu arteritis is a troubling problem because it affects women of childbearing age, and it will cause risks to the mother and the child. The arteritis could be associated with Crohn's and rectocolitis disease [6].

	Classification	Diseases
1	Large vessel vasculitis	Takayasu arteritis Giant cell arteritis
2	Medium vessel	Polyarteritis nodosa Kawasaki disease
3	Small vessel vasculitis ANCA-associated vasculitis	Microscopic polyangiitis Granulomatosis with polyangiitis (Wegener's) Eosinophilic granulomatosis with polyangiitis (Churg-Strauss)
4	Small vessel vasculitis Immune complex	Antiglomerular basement membrane disease Cryoglobulinemic vasculitis (CV) Hypocomplementemic urticarial vasculitis (anti-C1q vasculitis) IgA vasculitis (Henoch-Schönlein)
5	Variable vessel vasculitis	Behçet's syndrome Cogan's syndrome
6	Single-organ vasculitis	Cutaneous leukocytoclastic angiitis Cutaneous arteritis Primary central nervous system vasculitis
7	Vasculitis associated with systemic disease	Lupus vasculitis Rheumatoid vasculitis Sarcoid vasculitis Others
8	Vasculitis associated with probable cause	Hepatitis C virus-associated CV Hepatitis B virus-associated vasculitis Syphilis-associated aortitis Drug-associated immune complex vasculitis Drug-associated ANCA-associated vasculitis Cancer-associated vasculitis Others

Table 1.
2012 International Chapel Hill Consensus Conference Nomenclature of Vasculitides [3].

The more effective treatments are oral corticosteroids, but there are many relapses when corticosteroids taper. High doses of corticosteroids (40–60 mg/day of prednisone or equivalent) should be initiated immediately after the diagnosis to induce remission. During disease remission, it is necessary sparing the drug until 15-20mg/day for two or three months, and after a year decrease the doses to 10mg/day or less. The immunosuppressive agents are used as corticoid sparing agents such as methotrexate, azathioprine, leflunomide, mycophenolate mofetil and cyclophosphamide [6]. There are evidences that biological agents such as anti-TNF-alpha, tocilizumab and rituximab could be effective in refractory cases, but more trials are necessary to evaluate the effectiveness of these drugs [8]. In addition, patients with ischemic problems need endovascular interventions. The mortality range is between 3 and 21% [6].

3.1.2 Giant cell arteritis (GCA)

It is a systemic vasculitis of large vessels, with tropism to the aorta and its branches, mainly external carotid. It is a most common vasculitis in above 50-year-olds and affects more women than men (3 W:2 M). The corticosteroids are the most common therapeutic used in these cases. In the beginning, we need high doses of corticosteroids to control the inflammation, and then it could be used sparingly until controlled in the patient. These strategies are valid for simple and complicated forms of the disease [7].

Simple form is defined by cephalic isolated symptoms with visual disturbance or changes in central nervous system. Nowadays, the gold standard treatment is oral prednisone or equivalent 40–60 mg/day, and methylprednisolone is not used anymore. The initial dose recommended is 0.7 mg/kg day of prednisone and the maximum dose is 80 mg/day [7].

The complicated form is defined when there is an ophthalmologic or a central nervous or extracephalic complication, mainly aorta and its branches. In these cases, methylprednisolone in pulses of 500 mg^{-1} g a day shall be used for 1 to 5 consecutive days. During maintenance, some authors believe in doses less than 5 mg/day. The American Society of Ophthalmology recommends pulses of intravenous methylprednisolone, when patients have ophthalmologic symptoms, but the Rheumatology Society prefers oral corticosteroid [7].

It is strategic to introduce corticosteroid-sparing drugs when the disease course is lasting long. Methotrexate is the most used treatment and with more evidence levels. There are four trials with methotrexate as a sparing drug, but the doses and the management are variable from 7.5–15 mg/week [7].

Cyclophosphamide has demonstrated effectiveness in patients dependent on corticosteroids or resistant to treatment. It is necessary to give more than 20 mg/day for 6 months or 10 mg/day for 1 year or more. The dose from 500 mg/m^2 or 500 mg/injection in 6 courses is standard from 5 months on average. It is highly important to warn patients of several side effects such as bone marrow suppression [7].

Azathioprine therapy has less controlled studies. Some studies showed a modest improvement. Hydroxychloroquine, an antimalarial synthetic drug, was tested also. One French study divided patients into two groups: one took prednisone 0.7 mg/kg/day in the beginning associated with hydroxychloroquine (400 mg/day) and the control group just received prednisone and placebo. The group that had hydroxychloroquine and corticoid had usually stopped later the prednisone and they had more relapses [7].

Nowadays, one option is biologics medicines. Anti-TNF-alpha does not have enough studies that showed control of the disease with this class of medication [7]. Tocilizumab, a humanized antibody that blocked membranous and soluble receptors of IL-6 (IL-6R), is a current option since IL-Il-6 is implicated in the etiopathogenesis of this affection. Three main studies evaluated the efficacy of the drug, although the therapeutics was different between these studies [7, 8].

A randomized control trial with 30 patients had tested 20 patients receiving corticoids and tocilizumab (8 mg/kg every 4 week during a year), and 10 received corticoids and placebo. The survival without relapses during a year was highest in patients treated with tocilizumab. However, there are no data available after the medication was stopped [7, 8].

Another promising biotherapy is abatacept. This medication with corticoids could decrease the risk of relapses, although more data is necessary to corroborate this hypothesis.

Ustekinumab, a subunit against anti-p40 IL-12/23 targeting Th1 and Th17 responses, has been showing similar side effects as corticoids. Some patients, who had refractory disease, have been treated with anakinra and they achieved success. Anakinra is a biopharmaceutical drug that blocks IL-1. All patients who had taken the drug had shown improvement in inflammation biomarkers and/or in their symptoms, with the disappearance of arterial inflammation in PET/CT. More studies are necessary though [7].

Other pharmacological drugs can be used as adjuvant treatments such as antiplatelets and anticoagulants, but there is no official recommendation about these drugs in the treatment of GAC. The statins do not influence the evolution of the disease, but they are used to prevent cardiovascular risk. Furthermore, these drugs could have an anti-inflammatory role via the inhibition of TH17 pathway [7].

3.2 Medium vessel vasculitis

3.2.1 Polyarteritis nodosa (PAN)

It is a rare necrotizing systemic vasculitis that affects small- and medium-sized vessels and is not usually associated with ANCA [8, 9], although there are reports in the literature of patients with PAN and positive ANCA. Several treatments are suggested for this condition, and the control is still a challenge [9].

Cutaneous involvement and peripheral nerves are the favorite sites of the disease, cutaneous and gastrointestinal vasculitis have specific histopathological characteristics, and until now, it has no development glomerulonephritis described. Gastrointestinal tract involvement is common and is one of the predictors of disease morbidity and mortality. There is cutaneous PAN without systemic involvement, and it very rarely progresses to the systemic form of the disease. According to the new classification, PAN is subdivided into idiopathic PAN and hepatitis B-associated PAN [9].

Treatment of PAN is usually based on the combination of systemic corticosteroids and immunosuppressants. The most commonly used medications are cyclophosphamide, azathioprine, methotrexate or mycophenolate mofetil [9].

The use of biological medications is reserved for cases of refractory PAN without association with hepatitis B. The use of rituximab, an anti-CD20 monoclonal antibody, has not been formally indicated for patients with PAN, but its use is supported in patients with ANCA-associated vasculitis. There are case reports using anti-TNF-alpha, such as etanercept and infliximab, and tocilizumab, but only in refractory cases [8].

3.2.2 Kawasaki disease (DK)

It is a systemic vasculitis, common in male child, with fever, rash, non-exudative bilateral conjunctivitis, oral and pharyngeal mucosal erythema, cervical lymphadenopathy, and it can affect the extremity. They may have fewer common symptoms such as pyuria, meningitis, shock and retropharyngeal or parapharyngeal abscess [10].

The etiology of Kawasaki disease is unknown. The diagnosis of Kawasaki disease is based on the presence of fever for ≥5 days, along with the presence of at least 4 of the 5 main clinical features [10] (**Table 2**).

	Clinical features
1	Erythema and cracking of the lips, strawberry tongue and/or pharyngeal and oral mucosa erythema
2	Bilateral bulbar conjunctival injection without exudate
3	Rash
4	Erythema and edema of the hands and feet in the acute phase and/or periungual scaling in the subacute phase
5	Cervical lymphadenopathy (≥1.5 cm in diameter), usually unilateral

Table 2.
Clinical features in Kawasaki disease [10].

Patients who have met the diagnostic criteria are considered to have complete Kawasaki disease (also referred to as typical or classic Kawasaki disease). Patients who do not have enough major clinical findings can be diagnosed with incomplete Kawasaki disease [10].

Intravenous immunoglobulin (IVIG) is the basis for treatment. Usually, it is initiated before 10 days of fever and significantly reduces coronary artery aneurysms (CAAs) that decrease from 25% to less than 5%. Around 10–30% of patients are resistant to IVIG treatment [10].

Unfortunately, in some cases, IVIG is discontinued or administered at a reduced dose due to cost. Evidence for therapies beyond IVIG is limited, with no evidence-based recommendations for the management of patients with resistance to initial IVIG treatment.

In the initial treatment, if the patient is in the acute phase treatment, aspirin (30–50 to 80–100 mg/kg/day) and IVIG (2 g/kg) are recommended. The treatment of patients with resistance (persistent or recurrent fever after the end of IVIG after 24 hours) is to initiate IVIG 2 g/kg, associated to corticosteroids and/or infliximab [10, 11].

The use of corticosteroids is controversial due to the cardiovascular risk in prolonged use.

Anti-TNF-alpha is also recommended for patients with coronary artery aneurysms, and in the minority of cases, it can be used in patients without aneurysms. Other therapies possible are anti-interleukin 1, canakinumab and cyclosporine [11].

3.3 Small vessel vasculitis (SVV)

3.3.1 Microscopic polyangiitis

Microscopic polyangiitis is an ANCA-associated vasculitis with significant morbidity and mortality. Treatment follows the same protocol as granulomatosis with polyangiitis [12].

3.3.2 Granulomatosis with polyangiitis (Wegener's)

It is another necrotizing systemic vasculitis that affects small and medium vessels and is often associated with ANCA. It occurs in patients between 45 and 60 years old from both genders and is rarely observed in Negroids. The main features are the involvement of the upper and lower respiratory tract and the kidneys. Ears, nose and throat may develop with sinusitis and crusted rhinorrhea that are usually severe. Pulmonary nodules and renal involvement with crescent glomerulonephritis can be seen [13, 14].

It is a serious disease, and if left untreated, it almost always progresses to death. With the advent of new therapies, 90% of patients evolve to remission and the survival rate is 80% in 10 years. The first phase, known as the induction phase, aims to put the patient into remission, and it lasts between 3 and 6 months according to the clinical response. The second phase is known as the maintenance phase trying to consolidate the first phase and prevent relapses. It lasts from 12–24 months [13] (**Figure 2**).

In induction, prednisone 1 mg/kg is recommended. For severe forms, methylprednisolone pulse is indicated at doses of 7.5–15 mg/kg/day for 1–3 consecutive days. After 3–4 weeks of treatment, the corticosteroid dose is gradually decreased, but without reaching doses of less than 15 mg/day until 4th month. The combination of two immunosuppressants in the induction phase is essential for severe or refractory patients such as cyclophosphamide or rituximab.

DOI: http://dx.doi.org/10.5772/intechopen.92778

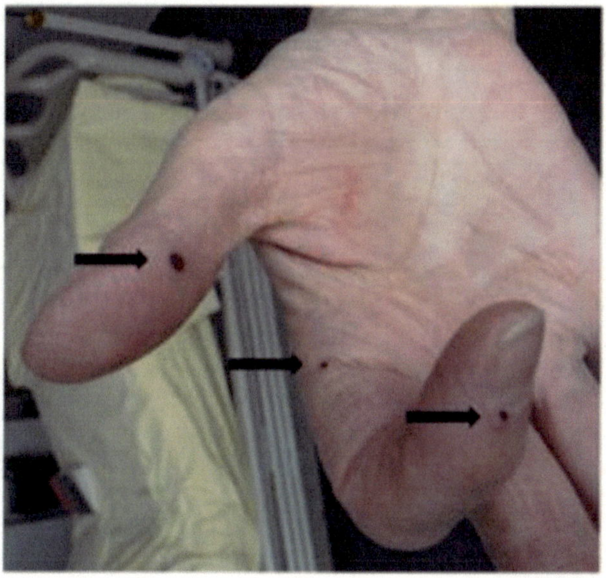

Figure 2.
Necrotic vascular purpura (black arrows) of the upper limbs in granulomatosis with polyangiitis-Wegener's [13].

Cyclophosphamide is preferred if rapid renal failure occurs at a dose of 600 mg/m^2 (maximum 1.2 g/bolus) every 2 weeks for 1 month (day+1, day+15, day+30) and then 700 mg/m^2 every 3 weeks until remission (average of 6–9 cycles in total).The dose can be adjusted for age and renal function (500 mg/m^2 in the presence of renal failure and 500 mg fixed dose every 3 weeks—maximum 6 bolus) [13, 14].

Rituximab is the choice for pregnant women or patients who have failed cyclophosphamide or have relapsed. It is used at a dose of 375 mg/m^2 per week for 4 weeks. Plasmapheresis may be used in severe forms of the disease with severe renal involvement (Cr > 500 μmol/L) or alveolar hemorrhage. Also, corticosteroid therapy can be associated with immunosuppressants [13, 14].

For localized or not very severe cases, methotrexate (20–25 mg per week) is an option. The AGATA study demonstrated efficacy of abatacept (1 mg/kg IV on day+1, day+29 and then once per month) combined with prednisone and an immunosuppressant (azathioprine, methotrexate or mycophenolate mofetil) for recurrent and limited forms.

Maintenance lasts between 18 and 24 months after remission has been achieved. Corticosteroids may be combined with azathioprine (2 mg/kg/day) or methotrexate (20–25 mg/week) [13].

Treatment with sulfamethoxazole/trimethoprim (400/800 mg) is given to prevent relapse from *Pneumocystis jirovecii* infection. Patients should be vaccinated following the regional schedule and is contraindicated live virus vaccines [13]. *Staphylococcus aureus* was the most identified pathogen in positive cultures, and it is important to remember that it also deserves prevention [14].

3.3.3 Eosinophilic granulomatosis with polyangiitis (Churg-Strauss)

Eosinophilic granulomatosis polyangiitis (EGPA), also known as Churg-Strauss disease, is characterized by patients with asthma, eosinophilia and necrotizing vasculitis with extravascular eosinophilic granulomas that affect small and large vessels. It is a vasculitis associated with ANCA (neutrophilic cytoplasmic antibodies) and the rarest vasculitis within the group of ANCA-associated vasculitis, and that is why there are no highly recommended treatments based on the literature [15].

Corticosteroids are usually used to induce remission of the disease as well as immunosuppressants, for example cyclophosphamide in more severe cases. Long-term maintenance of immunosuppressants is used to prevent disease recurrence, but their long-term efficacy is discussed. Azathioprine has been recommended as maintenance therapy. The efficacy of mepolizumab, an anti-IL5 monoclonal antibody, has been recommended for these patients alone or in combination with corticotherapy. Others monoclonal anti-IL5, reslizumab and benralizumab drugs are still being studied. Studies suggest that rituximab may be effective in EGPA. Other drugs such as IFN-α appear to be effective in remission induction and maintenance; however, the safety profile restricts their use. Drugs such as anti-IgE (omalizumab) are used to control asthma, but their effects are unknown in the treatment of vasculitis. High doses of immunoglobulin have also been used to induce disease remission with good results. Intravenous immunoglobulin may be effective in treating residual peripheral neuropathy [15].

3.4 Immune complex SVV

3.4.1 Cryoglobulinemic vasculitis

Cryoglobulins are cold precipitated immunoglobulins that can cause vasculitis and vasculopathy. Type I cryoglobulins are responsible for 10–15% of symptomatic vasculitis cases and are related to malignant hematological disorders such as myeloma, B-cell lymphoma or undetermined monoclonal gammopathy (MGUS). Mixed cryoglobulins correspond to 80–85% of cases and are associated with infectious diseases, especially chronic hepatitis C, B-cell malignancies and autoimmune diseases such as Sjögren's syndrome and lupus. Vasculitis is most associated with mixed cryoglobulins. Women are affected more than men 2:1 [12].

In symptomatic patients with cryoglobulinemia type I, it is indicated to treat the hematologic basis diseases. Lymphomas require a combination of chemotherapy and myeloma treatment with drugs such as bortezomib, thalidomide, lenalidomide and other alkylating agents. Bone marrow transplantation can be performed in patients with cryoglobulinemia-associated myeloma. MGUS can be treated with the same myeloma drugs; however, rituximab has been the drug of choice. Plasmapheresis may be used for patients with severe renal involvement or extensive lower limb necrosis. Avoiding exposure to cold is essential [12].

Mixed cryoglobulins, usually hepatitis C-associated cryoglobulinemic vasculitis, are well-treated as suppression of hepatitis C replication occurs. Studies reported between 2011 and 2013 associated with pegylated interferon (PegIFN) and ribavirin for 12 months achieved a 50–60% control response to hepatitis C. The introduction of antiviral agents dramatically changed cryoglobulinemia-associated vasculitis. They promote shorter treatment without the need for interferon and with responses greater than 95% associated with few adverse effects. The drugs used are sofosbuvir, simeprevir, ledipasvir and daclatasvir, that can be associated with ribavirin. In some cases of more severe vasculitis, low doses of rituximab and other immunosuppressants may be used in selected cases. Rituximab targets B-cell populations that produce cryoglobulins and treats severe vasculitis [12] (**Figure 3**).

3.4.2 Urticaria vasculitis

The term urticaria vasculitis is used for plaques of urticaria that present leukocytoclasia on histopathological examination. It is a clinical pathological diagnosis,

DOI: http://dx.doi.org/10.5772/intechopen.92778

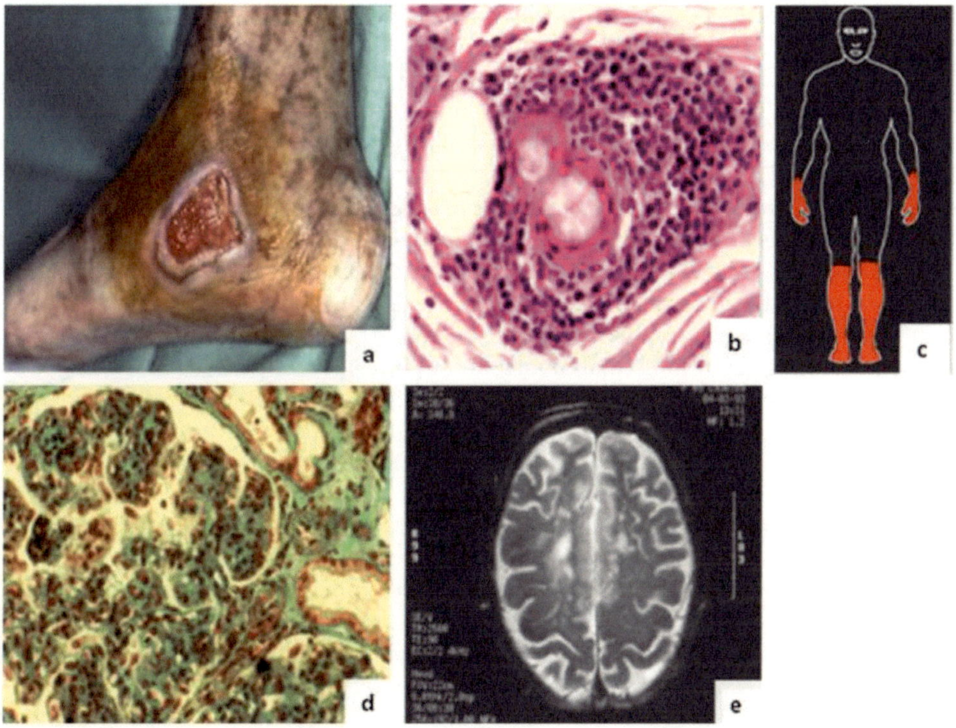

Figure 3.
Clinical manifestations of cryoglobulinemic vasculitis; (a) severe skin ulcer; (b) nerve biopsy specimen showing vasculitis with a perivascular inflammatory infiltrate; (c) distribution of the peripheral neurological involvement indicating length dependency; (d) renal biopsy showing membranoproliferative glomerulonephritis; (e) magnetic resonance imaging of the brain showing vasculitis [12].

characterized by a skin inflammation of the dermis capillaries and postcapillary venules, with a range of clinical signs from hives picture to a well-established vasculitis. Like any vasculitis, it can affect the skin and other organs, including skeletal, pulmonary, renal, gastrointestinal, cardiac and ophthalmic systems. The disease is spectral, ranging from mild to severe [16, 17].

Related etiological factors are infections, medications, autoimmune reactions, malignancies or idiopathic reasons. It can be classified as normocomplementemic or NUV that presents normal levels of complement, usually not with systemic involvement, and hypocomplementemic or HUV that has low complement levels and may have systemic involvement [17].

The treatment of vasculitis urticaria is a challenge, and there are no guidelines for the management of this disease. There are reports in the literature of the use of hydroxychloroquine and colchicine, which are as effective as systemic corticosteroids [16, 18].

Immunosuppressive drugs are used such as azathioprine, mycophenolate mofetil or cyclophosphamide. Rituximab with the usual doses also seems to have a good response. To manage the symptoms, sedative and nonsedating antihistamines may be used if urticaria is prominent. However, it is not usually effective. Nonsteroidal anti-inflammatory drugs are very useful in these cases such as naproxen, indomethacin and ibuprofen [13, 16].

Another alternative therapy is to combine corticosteroids up to 1 mg/kg and another immunosuppressant (methotrexate, mycophenolate mofetil, azathioprine, cyclosporine and very rarely cyclophosphamide). Recent reports include rituximab, anakinra, canakinumab and omalizumab as therapeutic arsenal [17].

3.4.3 Immunoglobulin A vasculitis (Henoch-Schönlein purpura)

It is the most common small vessel vasculitis of childhood, with a predominance of IgA deposits. Typically, it involves the skin, intestines and glomeruli and may be associated with arthralgia and/or arthritis. In most patients, only supportive treatment is required and analgesia [19, 20]. The disease regresses spontaneously most often within 4 weeks, but in some cases, it may last for more than 6 weeks [19]. In these cases, the therapeutic options may be dapsone (on average 1–2 mg/kg) or colchicine, but there are still no randomized controlled trials with the optimal dose and duration of treatment [18, 21]. Some patients require treatment with systemic corticosteroids, such as nephritis, orchitis, cerebral vasculitis, pulmonary hemorrhage and severe gastrointestinal involvement [19].

Other therapies used are the addition of cytotoxic immunosuppressants, intravenous immunoglobulins and plasmapheresis [20]. In case of severe rectal pain or abdominal pain, the use of systemic corticosteroids is recommended [19, 20].

The recommended doses for prednisolone are 1–2 mg/kg/day for 1–2 weeks followed by weaning. In severe cases, it is possible also to use methylprednisolone pulses 10–30 mg/kg maximum 1 g/day for 3 consecutive days. Corticosteroid prophylaxis to decrease the chance of developing nephritis is not indicated. There is evidence showing that the use of angiotensin-converting enzyme inhibitors (ACE inhibitors) may have beneficial effects in patients with proteinuria. In patients with proteinuria longer than 3 months, regardless of whether they are receiving prednisone or another immunosuppressive drug, the use of ACE inhibitors or angiotensin receptor blockers is recommended to prevent and/or limit secondary glomerular injury. The first choice for mild nephritis treatment is prednisolone. If proteinuria persists, azathioprine, mycophenolate mofetil, cyclosporine or other corticosteroid-sparing treatment can be started. Also, IV methylprednisolone pulses can be considered.

In the treatment for moderate nephritis, the first line is oral, or IV prednisolone, or IV methylprednisolone. Azathioprine, mycophenolate mofetil IV or cyclosporine may be used according to renal histopathological findings [19].

The treatment for severe nephritis however is high doses of corticosteroids and IV cyclosporine to induce remission, and low doses of corticosteroids associated with azathioprine or mycophenolate mofetil as maintenance treatment [19].

3.5 Variable vessel vasculitis

3.5.1 Behçet's syndrome

Behçet's syndrome has been known since ancient times and was described by Hippocrates, but it was reported as a separate disease by Huluci Behçet in 1937. It is currently classified as a vasculitis belonging to the subgroup of variable vessel vasculitis. It was initially reported in countries bordering the silk route, but nowadays, it is found all over the world. The most common manifestation is oral apotheosis and is seen in over 90% of cases [22].

It is characterized by progressing multiple attacks and remissions. The duration of remission may vary from one attack to another and from one system to another. The attack, in most cases, is followed by a complete recovery called Restitutio ad Integrum of the tissue. A good example is oral aphthosis, and it is exceptional to see a scar. In case of oral aphthosis, it is uncommon to have a scar. In other organs such as the eyes, central nervous system and vascular system, we usually observe sequelae, which may even progress to death. In these cases, aggressive treatment is mandatory [22].

Treatment will vary depending on systemic involvement. Mucocutaneous manifestations do not require aggressive treatment, only topical treatment. Arthritis goes on with outbreaks and remissions, which can last from weeks to months, but usually respond well to nonsteroidal anti-inflammatory drugs (NSAIDs). Usually, the remission in these cases is long and the patient may discontinue the medication. There is usually no progression to destruction and deformities [22, 23].

Treatment of the gastrointestinal tract depends on the severity. For other manifestations, aggressive treatments are often required, starting with cytotoxic/immunomodulatory drugs associated with corticotherapy. Eventually, biological drugs are used such as apremilast [23].

When the patient suffers from eye involvement, infliximab or adalimumab anti-TNF drugs may be considered as first- or second-line therapies, or in the exacerbation of pre-existing disease. The European League first recommends corticosteroid-associated azathioprine therapy in all patients with subsequent involvement and the addition of cyclosporine or infliximab, or switching to interferon alfa with or without corticosteroids in patients with severe involvement as more than two drop lines in visual acuity on a 10/10 scale and/or retinal disease (retinal vasculitis or macular involvement) [23]. Other biological drugs are being studied such as IL-1 and IL-6 blockers [8].

Intravitreal use of fluocinolone acetonide also has been suggested in refractory cases. In these cases, the increased intraocular pressure and infections should be monitored. With the higher cause of morbidity and mortality, it is vascular involvement. It can affect arteries and veins of any caliber, and with various presentations such as thrombosis, occlusions and aneurysms. Immunosuppressive therapy is one of choices in these cases and there is no consensus regarding anticoagulants [23].

A retrospective multicenter study demonstrated a decrease in relapse in immunosuppressant versus untreated patients. Behçet-associated Budd-Chiari syndrome has a high mortality rate, and monthly cyclophosphamide pulses associated with corticosteroid therapy are the treatment of choice. Anti-TNF-alpha was used in 5 patients with refractory disease. Two of these patients already had terminal liver disease during infliximab administration and died due to liver failure. Two patients were successfully treated with infliximab and the fifth patient was stable with etanercept but had dural sinus thrombosis during follow-up [23].

3.6 Single-organ vasculitis

3.6.1 Cutaneous leukocytoclastic angiitis

Cutaneous leukocytoclastic angiitis is an inflammation of small vessels, characterized by an inflammatory infiltrate associated with leukocytoclasia by neutrophil fragmentation and fibrinoid necrosis in small vessel postcapillary venules. It is the most common histological type of cutaneous vasculitis and usually is idiopathic, although antibiotics are also linked [24]. It is often clinically manifested by a palpable purpura, which is found anywhere on the body, but usually affects more lower limbs and can cause arthralgia [24, 25].

Several symptomatic treatments may be proposed to patients: analgesics, nonsteroidal anti-inflammatory drugs and antihistamines. For chronic or persistent vasculitis, dapsone and/or colchicine may be effective. The use of colchicine in the treatment is beneficial through its effect on reducing neutrophil chemotaxis, blocking leukocyte adhesion and stabilizing lysosomal membranes. Colchicine at a dose of 0.6–1.8 mg/day induces resolution within 1–2 weeks, according to several authors [18].

sometimes doses should be higher than usual due to resistance to anticoagulant effects. For treatment, as initial therapy, unfractionated heparin or low molecular weight heparin is used. Warfarin may also be used [28].

Because patients with APS and thrombosis are at high risk for recurrent thromboembolism episodes, prolonged oral anticoagulant therapy is the best option for attempting to prevent further episodes. The most used oral anticoagulant is warfarin with a therapeutic goal of maintaining INR greater than or equal to 3 [31].

In cases of APS secondary to the underlying systemic disease, it should not open the treatment with systemic oral corticoid. Other agents that can be used are plasmapheresis, immunoglobulin and dapsone, among others [27].

In refractory and catastrophic cases that there is multiple organ infarction, anticoagulation combinations, steroids, plasmapheresis, intravenous immunoglobulin and fish oil derivatives can be used. Fibrinolytic agents have no proven benefit [28].

In case of pregnancy, the best alternative during this period is heparin associated with low doses of aspirin. Combined treatment is more effective to prevent miscarriages than just aspirin alone. Unfractionated heparin, low molecular weight heparin (enoxaparin 40 mg/day) and dalteparin 5000 UI/day can be used during this period. Warfarin should not be used in pregnant women. Accidental discovery of antiphospholipid antibodies during pregnancy, with no clinical history of problems such as thromboembolic events or systemic lupus erythematosus, does not require treatment [28].

4.2 Blue finger syndrome

It is a sudden cutaneous manifestation in which the fingers, especially the toe, develop cyanotic and painful character. The priority etiology is embolic, but there may be other causes such as rheumatologic and neoplastic, among others. Treatment for this condition depends on the treatment of the underlying disease. However, general and local care such as limb warm-up, physical protection, treatment of secondary infections are essential [32].

4.3 Acrocyanosis

Acrocyanosis is a disease resulting from chronic vasospasm that causes reflex vasodilation in the affected extremities, usually by medication or central nervous system disorders. It may be painful, cold, discolored, hyperhidrosis, paresthesia and even tingling. Like for blue finger syndrome, treatment is only supportive [27].

4.4 Erythromelalgia

Erythromelalgia is characterized by vasodilation of the extremities, especially in male children, and is usually associated with limb warm-up, pain and burning. It is believed that there is some change in calcium channels, so therapy is directed toward this focus.

In the case of primary erythromelalgia, anesthetics, antiarrhythmics, anticonvulsants and even oral magnesium may be used. In the case of secondary erythromelalgia, besides the treatment already discussed, the underlying disease needs to be controlled [27].

Its treatment includes topical drugs like 5% lidocaine and 0.075% capsaicin. For oral medications, we have amitriptyline 10 mg/day, gabapentin 900–1800 mg/day, pregabalin 75 mg/day, flecainide 200 mg/day and buflomedil 200–330 mg/day. Tricyclic antidepressants, selective serotonin reuptake inhibitors and, in selected cases, acetylsalicylic acid, beta blockers and calcium channel antagonists may be excellent associations. In refractory cases, we may use epidural infusions of opioids, bupivacaine and, in the latter case, sympathectomy. However, responses are quite variable and complete remission of symptoms is rarely observed [27, 33].

4.5 Livedo reticularis

Reticular livedo is a common dermatological manifestation in which the limb in question suffers a vasospasm and has a cyanotic, erythematous and erythematous-violet coloration. When the blood plot does not have a confluent pattern, it is called a racemose livedo. In clinical practice, it may be isolated by cold or trauma or may be associated with some systemic diseases such as lupus erythematosus, scleroderma and HIV. Although it is more common in the limbs, it can also affect the trunk and there may be ulcerations. Cold stimulates vasospasm, when the cause is removed, however, over time, vessels may become permanently dilated and become permanently telangiectatic [27, 34].

The treatment of livedo, primarily, is protection against the cold. Vasodilators may be an alternative and corticosteroids should be avoided as much as possible. If there are ulcerations or a racemose livedo associated with antiphospholipid antibody syndrome, anticoagulation is recommended. When livedo is associated with some underlying disease, the management of the disease usually improves its manifestation. Other medications may be tried such as danazol, tissue plasminogen activator (tPA), pentoxifylline and antiplatelets. Immunosuppressants such as azathioprine and sympathectomy are reserved for refractory cases [27, 34, 35].

4.6 Raynaud's phenomenon

It is a paroxysmal vasospastic disorder characterized by the simultaneous alternation of pallor (vasoconstriction), cyanosis (blood stasis) and redness (compensatory vasodilation).

We consider Raynaud's disease (20% of the cases) when it occurs primarily and not associated with other acne. And we consider Raynaud's phenomenon (80%) when it occurs secondary to another disease [27].

Treatment consists of treating the basic disease and quitting smoking. Protection against the cold is necessary to avoid triggering the frame. In idiopathic forms, the use of nifedipine 30–120 mg/day can be used. Other treatment modalities may also be employed such as topical nitroglycerin, iloprost, losartan, serotonin receptor inhibitors, phosphodiesterase inhibitors, n-acetylcysteine, botulinum toxin, bosentan, platelet inhibitors and fibrinolytics. In refractory cases, sympathectomy may be performed [27].

4.7 Erythema pernio or perniosis

Perniosis is a rare panniculitis that develops with painful erythematous-violaceous nodules in young people more susceptible to cold. There is also a certain uncertain relationship with tobacco. Treatment, like other cold-related conditions, requires protection from the cold such as appropriate clothing, gloves, socks, boots and smoking cessation. Behavior changes and topical corticoid creams can help to heal the lesions. Some vasodilators may also be applied as nicotinic acid, nifedipine and pentoxifylline. In refractory cases, sympathectomy and UVB phototherapy may also be considered [36].

4.8 Livedoid vasculitis or Millian's white atrophy

Livedoid vasculitis is one of the most common vasculopathies, described in 1929, and can affect up to about 5% of the healthy population, reaching 70% in patients with venous ulcers. Some authors differentiate livedoid vasculitis from Millian's white atrophy in relation to etiology. Livedoid vasculitis is persistent livedoid reticularis associated with an ulcer [37]. White atrophy is a white atrophy scar with a stellar pattern most common in the lower limbs of women aged 30–60 years that may or may not be associated with collagenases and neoplasia; however, the best-established link is with chronic venous insufficiency [35].

The treatment for this disease is divided into some of the following groups:

- Vasodilators such as nifedipine [27, 35].

- Drugs stimulating endogenous fibrinolytic activity such as danazol, the activating factor of recombinant tissue plasminogen, rt-PA, low dose alteplase, associated or not with heparin and aspirin [27, 35].

- Drugs such as dipyridamole, cilostazol, the thienopyridine group (clopidogrel, ticlopidine hydrochloride, whether associated with aspirin) and sarpogrelate.

- Hemorheological drugs that decrease blood viscosity, increase red blood cell flexibility and improve circulation. From this group, the example is pentoxifylline and buflomedil hydrochloride [27, 35].

- Modulating lymphocyte response therapy as systemic PUVA phototherapy [27].

- Other drugs such oral corticosteroids, intravenous immunoglobulin, cyclosporine, hyperbaric oxygen therapy and intravenous iloprost [27].

For patients with any associated thrombophilia, warfarin, unfractionated heparin, low molecular weight heparin and even heparin minidoses may also be used [27].

4.9 Atherosclerosis obliterans

It is a chronic disease associated with diabetes that gradually occludes the vessel light. It affects the feet more than the hands in patients over 50 years. There is no specific treatment, but oral corticosteroids may be employed associated with diabetes with strict disease control. Behavioral change with diabetes mellitus improvement and antibiotic therapy, if secondary injuries, is the therapy of choice. Vasodilators do not have their proven efficiency. In severe cases, sympathectomy and vascular surgery may be used [38].

4.10 Thromboangiitis obliterans (Buerger's disease)

The priority treatment is smoking cessation and rest. Iloprost IV may be an alternative to retard evolution. In many cases, amputation is required, with excision of the gangrenous fingers, sympathectomy and surgeries such as arterialization of the venous arch of the foot. The vascular surgeon is extremely important for follow-up [27, 39].

4.11 Malignant atrophic papulosis (Köhlmeier-Degos disease)

Malignant atrophic papulosis (MAP) is an obliterating endovasculitis of small- and medium-sized arteries that produces tissue infarction as its main feature. It is considered an uncommon disease of unknown cause and can affect the skin, gastrointestinal tract and central nervous system, and the involvement of these last two systems can be fatal [40].

There is no fully effective treatment for the disease. Some authors use acetylsalicylic acid (300 mg daily) and/or dipyridamole (75 mg twice daily) as the first therapeutic modality, which facilitates blood perfusion. Other therapeutic options such as aspirin, heparin and warfarin can be used, however, aspirin is more associated with resurgence of lesions when discontinuing the drug [41].

More recently, studies have been conducted using eculizumab that have shown initial efficacy in skin and intestinal lesions, but the drug has not been able to prevent the development or progression of systemic manifestations. Subcutaneous treprostinil has been successfully tested in some cases with dramatic and sustained improvement in clinical status, although the response was not immediate. The mechanism of action of treprostinil in this scenario is not yet well understood [42, 43].

The use of corticosteroids, chloroquine or other immunosuppressants has proved unsatisfactory and has great potential to worsen the disease by unknown mechanism; therefore, they are not indicated [43].

4.12 Superficial thrombophlebitis

These are vascular inflammations with thrombus formation and consequent occlusion or may occur due to slow flow within a varicose vein. If thrombophlebitis is found in apparently normal superficial veins, attention should be paid to the possibility of underlying malignancy, thrombosing coagulopathy and silent deep vein thrombosis [44].

For therapeutic management, in cases of limited superficial thrombophlebitis below the knee, without evidence of deep vein thrombosis, compression by specific stockings and the use of nonsteroidal anti-inflammatory drugs are enough, providing symptomatic relief. However, if there is deep venous thrombosis or extension to the saphenofemoral or saphenopopliteal junctions, prophylactic use of low molecular weight heparin may be necessary [44].

5. Conclusion

Vasculitis and vasculopathies are a challenge physicians face on a daily basis. Due to rarity of the diseases all over the globe, the scientific community is not able to perform studies with a great number of patients and biological medications seem to be the promise to a cure or the disease control. These groups of drugs are relatively new and still expensive in most countries. Furthermore, more studies need to be developed and long follow-ups should be performed before they are considered gold-standard treatment. In the medical reality nowadays, despite any consideration, an early diagnosis can change the whole disease course and prevent inabilities, and even without cure, it is indispensable to control symptoms and provide a better quality of life to patients.

DOI: http://dx.doi.org/10.5772/intechopen.92778

Author details

Ana Paula Urzedo[1,2]*, Mariane Martins Manso[1,3,4] and Thaisa Macarini de Faria[5,6,7]

1 Dermatology, BWS Institute, Brazil

2 Cosmiatry, Faculdade de Medicina do ABC and Hospital Albert Einstein, Brazil

3 Faculdade de Medicina do ABC, Brazil

4 Health Administration, Fundação Getulio Vargas, Brazil

5 Faculdades Integradas da União Educacional do Planalto Central, Brazil

6 Clinical Dermatology, Instituto BWS, Brazil

7 Cosmiatry, Instituto Superior de Medicina, Brazil

*Address all correspondence to: apurzedo.71@gmail.com

References

[1] Radic M, Radic J. Updates in the Diagnosis and Treatment of Vasculitis. London: IntechOpen; 2013. Available from: https://www.intechopen.com/books/updates-in-the-diagnosis-and-treatment-of-vasculitis/vasculitis-and-vasculopathy-in-rheumatic-diseases

[2] Micheletti RG. Cutaneous vasculitis in rheumatologic disease: Current concepts of skin and systemic manifestations. Clinics in Dermatology. 2018;**36**(4):561-566. Available from: https://www.ncbi.nlm.nih.gov/pubmed/30047438 [Accessed: 13 July 2019]

[3] Caproni M, Verdelli A. An update on the nomenclature for cutaneous vasculitis. Current Opinion in Rheumatology. 2019;**31**(1):46-52. Available from: https://www.ncbi.nlm.nih.gov/pubmed/30394939 [Accessed: 17 July 2019]

[4] Ortiz-Sanjuán F, Blanco R, Hernández JL, Pina T, González-Vela MC, Fernández-Llaca H, et al. Drug-associated cutaneous vasculitis: Study of 239 patients from a single referral center. The Journal of Rheumatology. 2014;**41**(11):2201-2207. Available from: https://www.ncbi.nlm.nih.gov/pubmed/25225278 [Accessed: 17 July 2019]

[5] Sunderkötter CH, Zelger B, Chen KR, Requena L, Piette W, Carlson JA, et al. Nomenclature of cutaneous vasculitis: Dermatologic addendum to the 2012 revised International Chapel Hill Consensus Conference Nomenclature of Vasculitides. Journal of the German Society of Dermatology. 2018;**70**(2):171-184. Available from: https://www.ncbi.nlm.nih.gov/pubmed/29136340 [Accessed: 13 July 2019]

[6] Seyahi E. Takayasu arteritis: An update. Current Opinion in Rheumatology. 2017;**29**(1):51-56. Available from: https://www.ncbi.nlm.nih.gov/pubmed/27748689 [Accessed: 17 July 2019]

[7] Watelet B, Samson M, de Boysson H, Bienvenu B. Treatment of giant-cell arteritis, a literature review. Modern Rheumatology: Japanese Journal of Rheumatology. 2017;**27**(5):747-754. Available from: https://www.ncbi.nlm.nih.gov/pubmed/27919193 [Accessed: 13 July 2019]

[8] Krusche M, Ruffer N, Kötter I. Tocilizumab treatment in refractory polyarteritis nodosa: A case report and review of the literature. Rheumatology International. 2019;**39**(2):337-344. Available from: https://www.ncbi.nlm.nih.gov/pubmed/30465270 [Accessed: 13 July 2019]

[9] Alibaz-Oner F, Koster MJ, Crowson CS, Makol A, Ytterberg SR, Salvarani C, et al. Clinical spectrum of medium-sized vessel vasculitis. Arthritis Care & Research. 2017;**69**(6):884-891. Available from: https://www.ncbi.nlm.nih.gov/pubmed/27564269 [Accessed: 13 July 2019]

[10] Li Y, Yang Q, Yu X, Qiao H. A case of Kawasaki disease presenting with parotitis: A case report and literature review. Medicine. 2019;**98**(22):e15817. Available from: https://www.ncbi.nlm.nih.gov/pmc/articles/PMC6708832/ [Accessed: 13 July 2019]

[11] Dionne A, Burgner D, De Ferranti S, Singh-Grewal D, Newburger J, Dahdah N. Variation in the management of Kawasaki disease. Archives of Disease in Childhood (ADC). 2019:317191. Available from: https://www.ncbi.nlm.nih.gov/pubmed/31196912

[12] Desbois AC, Cacoub P, Saadoun D. Cryoglobulinemia: An update in 2019. Joint, Bone, Spine. 2019;**86**(6):707-713. Available from: https://www.ncbi.nlm.

nih.gov/pubmed/30731128 [Accessed: 13 July 2019]

[13] Comarmond C, Cacoub P. Granulomatosis with polyangiitis (Wegener): Clinical aspects and treatment. Autoimmunity Reviews. 2014;**13**(11):1121-1125. Available from: https://www.ncbi.nlm.nih.gov/pubmed/25149391. [Accessed: 13 July 2019]

[14] Jennette JC, Nachman PH. ANCA glomerulonephritis and vasculitis. Clinical Journal of the American Society of Nephrology. 2017;**12**(10):1680-1691. Available from: https://www.ncbi.nlm.nih.gov/pubmed/28842398 [Accessed: 13 July 2019]

[15] Furuta S, Iwamoto T, Nakajima H. Update on eosinophilic granulomatosis with polyangiitis. Allergology International. 2019;**68**(4):430-436. Available from: https://www.ncbi.nlm.nih.gov/pubmed/31266709 [Accessed: 13 July 2019]

[16] Younger DS, Carlson A. Dermatologic aspects of systemic vasculitis. Neurologic Clinics. 2019;**37**(2):465-473. Available from: https://www.ncbi.nlm.nih.gov/pubmed/30952419 [Accessed: 13 July 2019]

[17] Davis MDP, Hilst, JCVD. Mimickers of urticaria: Urticarial vasculitis and autoinflammatory diseases. The Journal of Allergy and Clinical Immunology: In Practice. 2018;**6**(4):1162-1170. Available from: https://www.ncbi.nlm.nih.gov/pubmed/29871797 [Accessed: 13 July 2019]

[18] Robinson KP, Chan JJ. Colchicine in dermatology: A review. Australasian Journal of Dermatology. 2018;**59**(4):278-285. Available from: https://www.ncbi.nlm.nih.gov/pubmed/29430631. [Accessed: 17 July 2019]

[19] Ozen S, Marks SD, Brogan P, Groot N, de Graeff N, Avcin T, et al. European consensus-based recommendations for diagnosis and treatment of immunoglobulin A vasculitis: The SHARE initiative. Rheumatology. 2019;**58**(9):1607-1616. Available from: https://www.ncbi.nlm.nih.gov/pubmed/30879080 [Accessed: 13 July 2019]

[20] Mauro A, Mauro S, Rega R, Martemucci L, Sottile R. Successful treatment of hemorrhagic bullous Henoch-Schonlein purpura with intravenous immunoglobulins. Pediatric Dermatology. 2019;**36**(1):e34-e36. Available from: https://www.ncbi.nlm.nih.gov/pubmed/30561101 [Accessed: 13 July 2019]

[21] Roman C, Dima B, Muyshont L, Schurmans T, Gilliaux O. Indications and efficiency of dapsone in IgA vasculitis (Henoch-Schonlein purpura): Case series and a review of the literature. European Journal of Pediatrics. 2019;**178**(8):1275-1281. Available from: https://www.ncbi.nlm.nih.gov/pubmed/31230197 [Accessed: 13 July 2019]

[22] Davatchi F. Behçet's disease. International Journal of Rheumatic Diseases. 2018;**21**(12):2057-2058. Available from: https://www.ncbi.nlm.nih.gov/pubmed/30681276 [Accessed: 17 July 2019]

[23] Ozguler Y, Hatemi G. Management of Behçet's syndrome. Current Opinion in Rheumatology. 2016;**28**(1):45-50. Available from: https://www.ncbi.nlm.nih.gov/pubmed/26555450 [Accessed: 17 July 2019]

[24] Cakiter AU, Kucuk OS, Ozkaya DB, Topukcu B, Onsun N. Demographic characteristics, aetiology, and assessment of treatment options in leukocytoclastic vasculitis. Postepy Dermatologii I Alergologii. 2017;**34**(2):104-109. Available from: https://www.ncbi.nlm.nih.gov/pmc/articles/PMC5420600/ [Accessed: 13 July 2019]

[25] Gambichler T, Kulik MA, Skrygan M, Rooms I, Höxtermann S. Cutaneous leukocytoclastic vasculitis: The role of lymphocytes and related immune markers. Postępy Dermatologii I Alergologii. 2017;**34**(4):299-305. Available from: https://www.ncbi.nlm.nih.gov/pmc/articles/PMC5560176/ [Accessed: 11 June 2019]

[26] Aounallah A, Arouss A, Ghariani N, Saidi W, Sriha B, Denguezli M. Cutaneous leukocytoclastic vasculitis: About 85 cases. The Pan African Medical Journal. 2017;**26**:138. Available from: https://www.ncbi.nlm.nih.gov/pubmed/28533861 [Accessed: 17 July 2019]

[27] Rivitti EA. Dermatologia: De Sampaio e Rivitti. 4ª ed. São Paulo: Artes Médicas; 2018:527-564

[28] Santamaria JR, Badziak D, Barros MFD. Síndrome antifosfolípide. Anais Brasileiros de Dermatologia. 2005;**80**(3):225-239. Available from: http://www.scielo.br/pdf/abd/v80n3/v80n3a02.pdf [Accessed: 17 July 2019]

[29] Dutra LA, Braga-Neto P, Pedroso JL, Barsottini OGP. Sneddon's syndrome: Case report and review of its relationship with antiphospholipid syndrome. Revista Einstein. 2012;**10**(2):230-232. Available from: https://journal.einstein.br/pt-br/article/sindrome-de-sneddon-relato-de-caso-e-revisao-sobre-a-relacao-com-a-sindrome-do-anticorpo-antifosfolipidio/ [Accessed: 12 July 2019]

[30] Brito FA, Tófani MLMC, Tófani FA, Kakehasi AM, Lanna CCD, Carvalho MAP. Libmann-sacks endocarditis and oral anticoagulation. Arquivos Brasileiros de Cardiologia. 2004;**82**(4):378-380. Available from: http://www.scielo.br/scielo.php?pid=S0066-782X2004000400009&script=sci_arttext [Accessed: 12 July 2019]

[31] Criado PR, Rivitti EA, Vasconcellos C, Valente NYS, Martins JEC. Cutaneous manifestations of thrombophilia. Anais Brasileiros de Dermatologia. 2008;**83**(6):491-506. Available from: https://core.ac.uk/download/pdf/37453598.pdf [Accessed: 17 July 2019]

[32] Reis PEO. Síndrome do dedo azul. Jornal Vascular Brasileiro. 2005;**4**(4):391-393. Available from: http://www.scielo.br/pdf/jvb/v4n4/v4n4a16.pdf [Accessed: 17 July 2019]

[33] Leroux MB, Carbo E. Revisiones Terapéuticas: Manejo racional de la eritromelalgia. Actualizaciones Terapeútica Dermatologicas. 2010;**33**:126. Available from: https://www.researchgate.net/publication/306392597_REVISIONES_TERAPEUTICAS_Manejo_racional_de_la_eritromelalgia_Manejo_racional_de_la_eritromelalgia [Accessed: 17 July 2019]

[34] Criado PR, Faillace C, Magalhães LS, Brito K, de Carvalho JF. Livedo reticularis: classification, causes e differential diagnosis. Acta Reumatologica Portuguesa. 2012;**37**(3):218-225. Available from: https://www.researchgate.net/profile/Paulo_Criado/publication/235004250_Livedo_reticularis_Classification_causes_and_differential_diagnosis/links/02bfe5103d0eb1b4f5000000/Livedo-reticularis-Classification-causes-and-differential-diagnosis.pdf [Accessed: 17 July 2019]

[35] Wolff K, Goldsmith LA, Katz SI, Gilchrest BA, Paller AS, Leffell DJ. Fitzpatrick's: Dermatology in General Medicine. 7th ed. United States of America: McGraw Hill Medical; 2008;**88**:1603-1604

[36] Tonoli RE, Souzai PRM. Case for diagnosis. Anais Brasileiros de Dermatologia. 2012;**87**(4):649-650.

Available from: http://www.scielo.br/scielo.php?script=sci_arttext&pid=S0365-05962012000400027 [Accessed: 17 July 2019]

[37] Zanini M, Wulkan C, Bertino D, Ito L. Livedoid vasculopathy or vasculitis? Anais Brasileiros de Dermatologia. 2003;**78**(6):755-757. Available from: http://www.scielo.br/scielo.php?script=sci_arttext&pid=S0365-05962003000600012 [Accessed: 17 July 2019]

[38] Pitta GBB, Castro AA, Burihan E. Angiologia e cirurgia vascular: guia ilustrado. 2nd ed. Maceió: UNCISAL/ECMAL & LAVA; 2003:1-21

[39] Busato CR, Utrabo CAL, Gomes RZ, Housome JK, Hoeldtke E, Pinto CT, et al. Arterialization of venous arch of the foot for the treatment of thromboangiitis obliterans. Jornal Vascular Brasileiro. 2008;7(3):267-271. Available from: http://www.scielo.br/pdf/jvb/v7n3/v7n3a13 [Accessed: 17 July 2019]

[40] Slaviero F, Annes RD, Frighetto L, Schirmer LM, Vanzin JR, Fröhlich AC, et al. Köhlmeier-Degos disease (malignant atrophic papulosis) and neurologic involvement. Arquivos de Neuro-Psiquiatria. 2009;**67**(3):692-694. Available from: http://www.scielo.br/scielo.php?script=sci_arttext&pid=S0004-282X2009000400022 [Accessed: 17 July 2019]

[41] Carvalho JFD. Degos' disease as a cutaneous manifestation of systemic lupus erythematosus. Acta Reumatológica Portuguesa. 2009;**34**(2A):246-250. Available from: http://www.actareumatologica.pt/repositorio/pdf/2A-09.pdf#page=91 [Accessed: 17 July 2019]

[42] Shapiro LS, Toledo-Garcia AE, Farrell JF. Effective treatment of malignant atrophic papulosis (Köhlmeier-Degos disease) with treprostinil: Early experience. Orphanet Journal of Rare Diseases. 2013;**8**:52. DOI: 10.1186/1750-1172-8-52 [Accessed: 17 July 2019]

[43] Theodoridis A, Makrantonaki E, Zouboulis CC. Malignant atrophic papulosis (Köhlmeier-Degos disease): A review. Orphanet Journal of Rare Diseases. 2013;**8**:10. Available from: https://www.ncbi.nlm.nih.gov/pmc/articles/PMC3566938/ [Accessed: 17 July 2019]

[44] Nasr H, Scriven JM. Superficial thrombophlebitis (superficial venous thrombosis). BMJ. 2015;**350**:h2039. Available from: https://www.bmj.com/content/350/bmj.h2039 [Accessed: 17 July 2019]

Chapter 10

Treatment of Vasculitis: Beyond the Basics

Muhammad Ishaq Ghauri and Muhammad Shariq Mukarram

Abstract

Vasculitis is the inflammation of blood vessels in the human body. It causes changes and remodeling in the walls of the vessels that include thickening, narrowing and scarring. As a result, the blood flow to the organs and tissues gets restricted leading to organ damage. The cause of primary vasculitis is not known; however, most cases are thought to be autoimmune. In the present era, it is getting difficult to treat vasculitis with conventional therapies, which includes cyclophosphamide, methotrexate, azathioprine and mycophenolate mofetil, with increasing rates of relapses. Since ever, corticosteroids and cytotoxic agents or immunosuppressants have been the mainstay for treating systemic vasculitis. However, the introduction of newer biological agents have bring about a revolution in the treatment of relapses and in cases where there is failure to induce and sustain remission.

Keywords: vasculitis, granulomatosis with polyangiitis, ANCA-associated vasculitis, giant cell arteritis, Takayasu arteritis, microscopic polyangiitis, eosinophilic granulomatosis with polyangiitis, anti-TNF alpha, monoclonal antibody, rheumatoid arthritis

1. Introduction

Vasculitis is a group of heterogeneous disorders that are characterized by inflammation, also sometimes necrosis of blood vessels that includes the veins, arteries and capillaries. Several different forms have been identified. The pathophysiology of vasculitis mainly involves the immune system of the body. In large vessel disease, particularly giant cell arteritis, it is a T cell-driven process activating the CD4 T cells which in turn promote the recruitment of macrophages and monocytes to the vessel wall causing vascular injury. This leads to release of various inflammatory markers and cytokines for example, Interleukin 1 and Interleukin 6, causing systemic inflammation [1]. Antineutrophil cytoplasmic antibody (ANCA)-associated vasculitis involves the activation of neutrophils that release inflammatory cytokines. They also induce formation of neutrophil extracellular traps that are necessary constituent of innate immunity. These neutrophil traps are injurious to small vessels that not only cause vascular injury but also produce antineutrophil cytoplasmic antibody, therefore producing a vicious cycle [2].

The aim of this chapter is to highlight the rising number of therapeutic options available to treat systemic vasculitis. Use of biologics has shown promising results, especially Rituximab and Infliximab while others remain in the pipeline. The recent emergence of these agents, that selectively targets the components of immune system, have brought an insurgence in treatment of systemic vasculitis. In this chapter we discuss the treatment of vasculitis beyond the prototype drugs (cyclophosphamide, azathioprine, mycophenolate mofetil, etc.), with biological agents.

2. Classification

American College of Rheumatology (ACR) presented the classification criteria for vasculitis in 1990 (**Table 1**). According to this criteria vasculitis was classified into primary and secondary types. Both these types were dependent on the size of vessel involved. Vasculitis affecting the large arteries include giant cell arteritis (GCA) and Takayasu arteritis. Medium vessel vasculitis includes polyarteritis nodosa (PAN) and Kawasaki disease, while small vessel vasculitis contains granulomatosis with polyangiitis (GPA) formerly known as Wegener's granulomatosis (WG), Churg-Strauss syndrome now known as eosinophilic granulomatosis with polyangiitis (EGPS), microscopic polyangiitis (MPA), Henoch Schonlein purpura and cryoglobulinemia. Secondary vasculitides encompass vasculitis secondary to rheumatoid arthritis and various infections (bacterial, viral and fungal) [3].

Infections affecting large arteries commonly include *Staphylococcus*, *Salmonella*, *Streptococcus*, coccidioidomycosis and treponema pallidum. Hepatitis B and C virus along with human immunodeficiency virus and Parvovirus B19 commonly affects medium sized vessels. Possible mechanism for infection related vasculitis includes (a) direct microbial invasion and (b) immune-mediated process either by humoral or cellular responses. Hepatitis B virus has firmly been seen with polyarteritis nodosa for more than 30 years. In France, the declining rate of hepatitis B infection has correlated with falling levels of hepatitis B-associated polyarteritis nodosa. There is a strong association between hepatitis C virus and mixed essential cryoglobulinemia. Vasculitis has been identified as a rare manifestation of human immunodeficiency virus. Multiple patterns have been described including polyarteritis nodosa, hypersensitivity vasculitis and large vessel disease [3].

Course of rheumatoid arthritis can be complicated by medium to small vessel vasculitis. This is common in men and linked with positive RA factor. It may present in a variety of ways including distal arteritis, splinter hemorrhages, peripheral neuropathy with mononeuritis multiplex and aortitis [4].

In year 2012 the International Chapel Hill Consensus conference adopted names for vasculitis on nomenclature of vasculitides as shown in **Table 2**. This classified vasculitis not only according to the size of vessel involved (as in ACR criteria) but also included a few other subtypes mainly, variable vessel vasculitis, single organ vasculitis, vasculitis associated with systemic diseases and vasculitis associated with probable etiology [5].

Dominant vessel	Primary	Secondary
Large arteries	Giant cell arteritis Takayasu arteritis	Aortitis associated with rheumatoid arthritis, infections
Medium arteries	Classic polyarteritis nodosa Kawasaki disease	Hepatitis B-associated, polyarteritis nodosa
Small vessels and medium arteries	Granulomatosis with polyangiitis, Churg-Strauss syndrome, microscopic polyangiitis	Vasculitis secondary to rheumatoid arthritis, drugs
Small vessels	Henoch-Schonlein purpura Cryoglobulinemia	Drugs, hepatitis C associated, infections

Table 1.
ACR classification of vasculitis [3].

Large vessel vasculitis
Takayasu's arteritis
Giant cell arteritis
Medium vessel vasculitis
Polyarteritis nodosa
Kawasaki disease
Small vessel vasculitis
Anti-neutrophil cytoplasmic antibody-associated vasculitis
Microscopic polyangiitis
Granulomatosis with polyangiitis
Eosinophilic granulomatosis with polyangiitis
Immune complex small vessel vasculitis
Cryoglobulinemic vasculitis
IgA vasculitis (Henoch Schonlein purpura)
Variable vessel vasculitis
Behcet's disease
Vasculitis associated with systemic disease
Lupus vasculitis
Rheumatoid vasculitis
Sarcoid vasculitis
Vasculitis associated with probable etiology
Hepatitis C-associated cryoglobulinemic vasculitis
Hepatitis B-associated vasculitis
Drug-associated immune complex vasculitis
Drug-associated ANCA-associated vasculitis

Table 2.
Modern Chapel Hill classification [5].

A wide range of drugs have been reported to cause a vasculitic reaction. ANCA-associated vasculitis has been attributed to use of various drugs including Hydralazine, Propylthiouracil, Allopurinol and Sulfasalazine. Leukotriene receptor antagonist, Montelukast and Zafirlukast, have been linked to Churg-Strauss syndrome. Acute vasculitis may be the presenting feature of an undiagnosed malignancy. Common malignancies associated with vasculitis are myelodysplasia, lymphoma and multiple myeloma [3].

3. Treatment with biological agents

3.1 Rituximab

It is a monoclonal antibody, which is directed against CD20 that is expressed on developing B cell. Although not clearly understood, this antibody can induce apoptosis of these developing B cells. In April 2011, Rituximab was the first agent to be approved by FDA for the treatment of vasculitis [6].

In the Rituximab in ANCA-associated vasculitis (RAVE) trial, patients with GPA and MPA who were being given steroid therapy were randomized and received either oral cyclophosphamide or I/V Rituximab (four infusions). Patients who were given cyclophosphamide were switched to azathioprine after going into remission while those with Rituximab were switched to oral placebo. At the end of 6 months, Rituximab was found to be non-inferior to cyclophosphamide at inducing remission and was found to be superior to cyclophosphamide for patients with relapsing disease [6].

Another study showed single course of Rituximab to be non-inferior to oral cyclophosphamide, which was followed by azathioprine for remission maintenance [7].

In the Rituximab versus cyclophosphamide in ANCA-associated vasculitis (RITUXVAS) trial, where 44 patients were diagnosed with ANCA-associated vasculitis, patients were randomized to receive Rituximab along with only two infusions of cyclophosphamide. This was compared with patients who were given I/V cyclophosphamide for 3–6 months followed by azathioprine. The rate of sustained remission was similar in both groups [8].

These studies indicate that Rituximab is comparable in efficacy to cyclophosphamide for remission induction. Moreover, Maintenance of remission under Rituximab in systemic ANCA-associated vasculitis (MAINRITSAN) trial 115 patients with either GPA or MPA were given either azathioprine or Rituximab in a dose of 500 mg IV × 2 doses (after achieving remission with cyclophosphamide). At the end of the study, major relapse rate was significantly lower in patients who received Rituximab [9].

Rituximab has also shown remarkable results in patients with EGPA whose disease was refractory to usual treatments (e.g. cyclophosphamide). One of the largest case series showed nine patients with EGPA refractory to conventional therapy, when treated with Rituximab were either in total or partial remission at the end of 3 months [10].

The safety and efficacy of Rituximab has been evaluated by Puechal et al. in patients with active systemic rheumatoid vasculitis (SRV). Out of 17 patients with active SRV who were treated with Rituximab, 12 patients achieved complete remission of their disease at the end of 6 months. The Birmingham Vasculitis Activity Score (BVAS) for rheumatoid arthritis dropped down from a baseline of 9.6 to 0.6 and the daily average dose of Prednisolone declined from 19.2 to 9.7 mg. After a year, 14 patients were in sustained remission [11].

Rituximab works by acting over B cells, clearing them from the body. It is the most extensively studied agent that has proved to be efficacious in most forms of vasculitis, especially ANCA-associated vasculitis. It is now the preferred choice over cyclophosphamide in order to reduce the adverse effect profile. Also, Rituximab is considered a suitable therapeutic option for inducing remission in patients with active vasculitis associated with rheumatoid arthritis.

Table 3 summarizes the clinical outcomes of different trials carried out using Rituximab as treatment for systemic vasculitis.

3.2 Infliximab

This is an IgG1, kappa monoclonal antibody specific for human tumor necrosis factor alpha. TNF alpha possesses multiple pro inflammatory properties for example, induction of Interleukin 1 and Interleukin 6, neutrophil activation etc. According to recent research, Infliximab is in phase 3 clinical development for treatment of Kawasaki disease [12].

Study	Agent	Disease	Outcome
RAVE trial	Rituximab	GPA and MPA	Rituximab was found to be non-inferior to cyclophosphamide at inducing remission and superior to cyclophosphamide for patients with relapsing disease
RITUXVAS trial	Rituximab	ANCA-associated vasculitis	Rate of sustained remission were similar in patients taking Rituximab vs. those given cyclophosphamide alone
MAINRTSAN trial	Rituximab	GPA/MPA	Significantly reduced the relapse rate

Table 3.
Rituximab trials [6, 8, 9].

One clinical trial studied the role of Infliximab in granulomatosis with polyangiitis exclusively. These patients were followed even after the discontinuation of Infliximab to monitor the remission maintenance. The reduction of their Birmingham Vasculitis Activity Score was significant. Surprisingly no severe adverse effects, deaths or infections were noted [13].

Use of TNF alpha inhibitors is not encouraged in giant cell arteritis. In a randomized controlled trial, some newly diagnosed patients with giant cell arteritis were given Infliximab along with corticosteroids (before tapering them). No significant difference was observed in patients who were successfully tapered off corticosteroids. Moreover, few subjects had a higher infection rate with the use of Infliximab [14].

Use of Infliximab has shown great effectiveness in refractory Kawasaki disease [15]. Single-dose Infliximab (5 mg/kg) was given to seven patients who failed to achieve remission with the standard therapy. These patients showed improvement without any adverse effects [16]. In another study, good response was seen in two patients with Kawasaki disease when treated with Infliximab who had a relapse with the conventional therapy [17].

This anti-TNF (tumor necrosis factor) agent has shown promising results in Wegener's granulomatosis in reducing the disease as well as inducing remission. Use of Infliximab is encouraged in medium vessel vasculitis. Patients with Kawasaki disease, who failed to respond to conventional therapy and those who experienced a relapse, reacted well to this agent. Unfortunately its use in giant cell arteritis is not promoted as suggested by a clinical study (as discussed above) in which patients with giant cell arteritis were treated with Infliximab showed no significant response, instead resulted in a higher rate of infection.

Table 4 summarizes the clinical outcomes of different studies and trials showing effectiveness of Infliximab in different types of vasculitis.

Study	Agent	Disease	Outcome
Lamprecht et al.	Infliximab	WG	Decrease in BVAS. No deaths, infections or adverse effects
Randomized trial	Infliximab	GCA	No significant difference. Increased rate of infection
Burns et al.	Infliximab	KD	Improvement in patients who failed to achieve remission by standard therapy/refractory disease

Table 4.
Infliximab trials [13–15].

3.3 Etanercept

This is one of the most rigorously studied agent, which is also a tumor necrosis factor inhibitor. Its role has been studied in GPA and MPA for maintenance of remission. In WGET, 174 patients received methotrexate or cyclophosphamide for their remission and were then randomized to get Etanercept or placebo. Unfortunately, there was no significant difference in rate of sustained remission between Etanercept and placebo [18].

Etanercept (25 mg twice weekly) was given to 20 patients with Wegener's granulomatosis over a period of 6 months, in twice-daily dose. Out of these patients, 70% had never had remission of their disease. This drug was combined with either cyclophosphamide or methotrexate. During the treatment, 80% patients went into disease remission and their Birmingham Vasculitis Activity Score fell significantly. However, three patients experienced major flare despite the therapy [19, 20].

The major drawback is the increased incidence of cancer in patients treated with Etanercept. Six of 92 patients developed a solid tumor. These tumors included mucinous adenocarcinoma of colon, metastatic cholangiocarcinoma, renal cell carcinoma and breast carcinoma [21].

Wegener's granulomatosis itself is also associated with increased risk of malignancy. The specific malignancies associated with it are bladder carcinoma, squamous cell carcinoma, leukemia and lymphomas [22].

Use of these two agents (Infliximab and Etanercept) have shown promising results in Takayasu arteritis as demonstrated by a case series in which 15 patients with treatment resistant disease were treated with either of the drug. After introduction of these agents, the average dose of corticosteroid dropped from 20 mg (range 12.5–40 mg) to 0 mg (range 0–20 mg). Among these 15 patients 93% showed remarkable improvement and 67% experienced steroid free remission for up to 3 years [23].

Etanercept is one of the most widely studied agents, which is also an anti-TNF drug, and has been seen to be beneficial not only in ANCA-associated vasculitis but also in large vessel vasculitis (Takayasu arteritis). The biggest disadvantage of this medication is the higher incidence of different types of cancers.

Table 5 summarizes two trials showing clinical outcomes of Etanercept in patients with Wegener's granulomatosis also known as granulomatosis with polyangiitis.

3.4 Belimumab

This is a human IgG1 gamma monoclonal antibody specific for soluble human B lymphocyte stimulator protein, also known as B cell-activating factor. Surprisingly, this is the only drug in late stage development for microscopic polyangiitis [12].

Currently this agent is in Phase 3 trial. Its efficacy and safety are being tested in a randomized, double blind study in combination with azathioprine. The dose given to patients is 10 mg/kg at days 0, 14, and 28 then after every 28 days till the study ends (clinical trials) [24].

Study	Agent	Disease	Outcome
WGET trial	Etanercept	WG	No significant difference in rate of sustained remission between Etanercept and placebo
Luqmani et al. and Stone et al.	Etanercept	WG	80% patients went into disease remission and their BVAS fell significantly. Three patients developed major flare

Table 5.
Etanercept trials [18–20].

Belimumab is a relatively newer agent that is currently under trial but has shown positive results in treatment of microscopic polyangiitis.

3.5 Mepolizumab

Mepolizumab is an Interleukin 5 humanized monoclonal antibody that binds to free Interleukin 5. It causes arrest of bone marrow eosinophil maturation. This monoclonal antibody is directed against Interleukin 5, which is a cytokine critical for activation of eosinophils. Mepolizumab when administered in a dose of 300 mg subcutaneously every 4 weeks, proved to be effective in prolonging disease remission, reducing the use of steroid [25].

Use of this agent has shown prompt normalization of peripheral eosinophil counts, as well as reduction in glucocorticoid usage. Two studies that showed use of Mepolizumab in EGPA, it led to decreased disease activity, normalization of eosinophil count and reduction of steroid use. However, cessation of this drug resulted in disease flare [26, 27].

Mepolizumab works by halting activation of eosinophils, acting directly on them. This biological agent is recommended in treating Churg-Strauss syndrome, although it is still under various trials. The major drawback is the disease flare caused after discontinuing the medication.

3.6 Tocilizumab

Tocilizumab is a humanized monoclonal antibody that binds to membrane-bound and soluble Interleukin 6 receptors and inhibits Interleukin 6 signaling pathways [28].

A study assessed eight patients who had refractory Takayasu arteritis. Two cases were refractory to Infliximab and three did not reach remission on steroids and methotrexate. Altogether eight patients received Tocilizumab and were followed for 18 months. Of these eight patients, seven achieved remission. This shows that Tocilizumab can be a potential therapy for patients with Takayasu arteritis refractory to anti-TNF alpha therapy [29].

A retrospective study assessed the effectiveness of Tocilizumab in complicated large vessel vasculitis. Patients were treated with Tocilizumab out of which eight had giant cell arteritis, two had large vessel vasculitis associated with rheumatoid arthritis and one had Takayasu arteritis. These patients were followed for 23 months. At the end of duration, seven patients were in remission, one patient relapsed after discontinuing the drug, and one patient suffered from serious infective complication. Two patients died, although cause of death was not attributable to the use of Tocilizumab. Three relapses occurred but remission was regained by switching the usual subcutaneous administration of Tocilizumab to intravenous [30].

Glucocorticoids are the conventional treatment for giant cell arteritis but adverse effects are common, so are the relapses, soon after tapering the steroids. Although the exact cause of death is not known, cytokines such as tumor necrosis factor alpha and Interleukin 6 have been implicated. A retrospective study included 134 patients from 40 different centers who were diagnosed with giant cell arteritis either by temporal artery biopsy or imaging techniques. All these patients had received high dose of steroids in past and majority of patients had been given biologic immunosuppressives such as Abatacept, Infliximab or Rituximab. Tocilizumab was given either intravenously (8 mg/kg 4 weeks apart) or subcutaneously (162 mg/week). At the end of 1 month the ESR and CRP had fallen and percentage of patients with anemia had decreased. Those who were followed for 2 years, amongst them 39 were seen

Study	Agent	Disease	Outcome
Mejla et al.	Tocilizumab	TA	Seven out of eight cases refractory to either Infliximab or methotrexate achieved remission
Toc in large vessel vasculitis (Marc Schmalzing)	Tocilizumab	GCA, RA vasculitis, TA	After follow up of 23 months, seven were in remission, one relapse after stopping the drug, three relapses but regained remission
IL-6 blocker exceeds (Nancy Walsh)	Tocilizumab	GCA	Patients who were on follow up till 2 years, out of them 39 were seen in remission. These patients had already taken biologics including Abatacept, Infliximab and Rituximab

Table 6.
Tocilizumab trials [29–31].

Agent	Mechanism of action	Dosage
Rituximab	Monoclonal antibody directed against CD20	500 mg intravenous
Infliximab	Anti-TNF alpha	5 mg/kg intravenous
Etanercept	Anti-TNF alpha	25 mg intravenous
Belimumab	Monoclonal antibody that inhibits B cell activating factor	10 mg/kg intravenous
Mepolizumab	Monoclonal antibody directed against Interleukin 5	300 mg subcutaneous
Tocilizumab	Monoclonal antibody directed against Interleukin 6	8 mg/kg intravenous 162 mg subcutaneous

Table 7.
Commonly used biological agents in the treatment of systemic vasculitis.

in remission with acute phase reactants within normal limits and minimum steroid dose (0–5 mg/day). However, after an average follow up of 12 months, 32 patients reported an adverse infection because of which 17 patients had to discontinue the therapy [31].

Tocilizumab works against the pro-inflammatory cytokine Interleukin 6 and has proven its efficacy in Takayasu arteritis that has failed to respond to Infliximab. Giant cell arteritis, non responsive to various other biological agents, has reacted remarkably to Tocilizumab by achieving disease remission in most of the cases. Despite all its applauding outcomes, life-threatening infection remains a serious complication.

Table 6 shows outcomes of studies that have evaluated the effectiveness of Tocilizumab in different types of vasculitis.

Table 7 summarizes the commonly used agents to treat systemic vasculitis, as discussed in this chapter, along with its mechanism of action and dosages.

4. Conclusion

In recent times use of high dose corticosteroids, cytotoxic and immunosuppressant drugs has improved the prognosis of systemic vasculitis dramatically. However, some patients still do not respond to conventional therapy or may not achieve remission. Few of them would relapse and a large number of patients develop illness secondary to the adverse effects caused by long term use of these drugs. The advent

DOI: http://dx.doi.org/10.5772/intechopen.92729

of biological agents has not just let to a better understanding of pathophysiology of systemic vasculitis, but has also proved to be safe and efficacious. Among these agents, anti-TNF and anti-B cell therapy have been the first choice in many cases. Although clinical data are still insufficient, these agents seem to occupy most of the market in near future [32].

5. Methods used for research of articles

We collected information by systematic review of the PubMed, scientific abstracts and by searching textbooks of Rheumatology.

Author details

Muhammad Ishaq Ghauri and Muhammad Shariq Mukarram*
Department of Medicine, Jinnah Medical College Hospital, Karachi, Pakistan

*Address all correspondence to: shariq.msm@gmail.com

References

[1] Guillevin L, Dorner T. Vasculitis: Mechanisms involved and clinical manifestations. Arthritis Research & Therapy. 2007;**9**(Suppl 2)

[2] Nakazawa D, Masuda S, Tomaru U, Ishizu A. Pathogenesis and therapeutic interventions for ANCA-associated vasculitis. Nature Reviews Rheumatology. 2019;**15**:91-101

[3] Mahr AD, Watts RA. Eular Textbook on Rheumatic Diseases. 1st ed. London: BMJ Group; 2012. p. 616

[4] Firestein GS et al. In: Erickson AR, Cannella AC, Mikuls TR, editors. Kelley & Firestein's Textbook of Rheumatology. 10th ed. Philadelphia, PA: Elsevier; 2017. p. 1181

[5] Firestein GS et al. In: Stone JH, editor. Kelley & Firestein's Textbook of Rheumatology. 10th ed. Philadelphia, PA: Elsevier; 2017. p. 1513

[6] Stone JH, Merkel PA, et al. Rituximab versus cyclophosphamide for ANCA-associated vasculitis. The New England Journal of Medicine. 2010;**363**:221-232

[7] Specks U, Merkel PA, et al. Efficacy of remission-induction regimens for ANCA associated vasculitis. The New England Journal of Medicine. 2013;**369**:417-427

[8] Jones RB, JWC T, et al. Rituximab versus cyclophosphamide in ANCA-associated renal vasculitis. The New England Journal of Medicine. 2010;**363**:211-220

[9] Guillevin L, Pagnoux C, et al. Rituximab versus azathioprine for maintenance in ANCA-associated vasculitis. The New England Journal of Medicine. 2014;**371**:1771-1780

[10] Thiel J, Hassler F, et al. Rituximab in the treatment of refractory or relapsing eosinophilic granulomatosis with polyangiitis (Churg Strauss syndrome). Arthritis Research & Therapy. 2013;**15**:R133

[11] Puechal X et al. Rituximab therapy for systemic vasculitis associated with rheumatoid arthritis: Results from the autoimmunity and rituximab registry. Arthritis Care and Research. 2012;**64**(3):331-339

[12] Fellner C. Biologics will pump up the vasculitis market. P&T Community. 2016;**41**(4):258-260

[13] Lamprecht P et al. Effectiveness of TNF alpha blockade with infliximab in refractory Wegener's granulomatosis. Rheumatology (Oxford). 2002;**41**:1303-1307

[14] Hoffman GS et al. Infliximab for maintenance of glucocorticosteroid-induced remission of giant cell arteritis. Annals of Internal Medicine. 2007;**146**:621-630

[15] Rodriguez M, Matamale MA, Segado AA. Infliximab as a novel therapy for refractory Kawasaki disease. Scandinavian Journal of Rheumatology. 2006;**35**:318-321

[16] Burns J et al. Immune monitoring in Kawasaki disease patients treated with infliximab and intravenous immunoglobulin. Clinical and Experimental Immunology. 2013;**174**: 337-344

[17] Burns JC et al. Infliximab treatment for refractory Kawasaki syndrome. The Journal of Pediatrics. 2005;**146**:662-667

[18] Wegener's Granulomatosis Etanercept Trial Research Group. Etanercept plus standard therapy for Wegener's granulomatosis. The New England Journal of Medicine. 2014;**371**:1771-1780

[19] Luqmani RA, Bacon PA, et al. Birmingham Vasculitis Activity Score (BVAS) in systemic necrotizing vasculitis. The Quarterly Journal of Medicine. 1994;**87**:671-678

[20] Stone JH, Hoffman GS, et al. A disease-specific activity index for Wegener's granulomatosis: Modification of the Birmingham Vasculitis Activity Score. Arthritis and Rheumatism. 2001;**44**:912-920

[21] Mukhtyar C, Luqmani R. Current state of TNF alpha blockade in Wegener's granulomatosis. Annals of the Rheumatic Diseases. Nov. 2005;**64**(Suppl 4):iv31-iv36

[22] Knight A, Askling J, Ekbom A. Cancer incidence in a population-based cohort of patients with Wegener's granulomatosis. International Journal of Cancer. 2002;**100**:82-85

[23] Katoh H et al. Takayasu's arteritis in a patient with Crohn's disease: An unexpected association during infliximab therapy. Internal Medicine. 2010;**49**(2):179-182

[24] Identifier NCT01663623, Belimumab in Remission of VASculitis (BREVAS). [Internet]. Bethesda (MD): National Library of Medicine (US); 18 April 2018. Available from: ClinicalTrials.gov. http://clinicaltrials.gov/ct/show/NCT01663623?

[25] Faverio P, Bonaiti G, Bini F, Pesci AVA. Mepolizumab as first targeted treatment for esinophilic granulomatosis with polyangiitis: A review of current evidence and potential place in therapy. Therapeutics and Clinical Risk Management. 2018;**14**:2385-2396

[26] Kim S et al. Mepolizumab as steroid sparing treatment options in patients with Churg Strauss syndrome. The Journal of Allergy and Clinical Immunology. 2010;**125**:1336-1343

[27] Moosig F et al. Targeting IL-5 in refractory and relapsing Churg-Strauss syndrome. Annals of Internal Medicine. 2011;**155**:341-343

[28] Hammoudeh M, Awadhi AA, Hasan EH, Akhlaghi M, Ahmadzadeh A, Abdollahi BS. Safety, tolerability and efficacy of tocilizumab in rheumatoid arthritis: An open-label phase 4 study in patients from the Middle East. International Journal of Rheumatology. 2015;**2015**:975028

[29] Mejla M et al. Efficacy and safety of interleukin 6 receptor monoclonal antibody (tocilizumab) in Colombian patients with Takayasu arteritis. Journal of Clinical Rheumatology. 2014;**20**:125-129

[30] Schmalzing M, Gadeholt O, Gernert M, Tony H-P, Schwaneck EC. Tocilizumab in large vessel vasculitis—Different routes of administration. Open Rheumatology Journal. 2018;**12**:152-159

[31] Walsh N. IL-6 blocker succeeds in real-world vasculitis—Clinical and laboratory improvements seen with tocilizumab in refractory giant cell arteritis. MedPage Today. 2019;**13**

[32] Pazzola G, Muratore F, Piptone N, et al. Biologics in vasculitides: Where do we stand, where do we go from now? Presse Médicale. 2015;**44**:e231-e239

Section 3

Methods for Diagnostic and Surgery in Microvascular Diseases

Chapter 11

The Impact of Nailfold Capillaroscopy in the Approach of Microcirculation

Vera Bernardino, Ana Rodrigues, Ana Lladó, Melissa Fernandes and António Panarra

Abstract

Nailfold capillaroscopy (NFC) is a simple, validated, and noninvasive method to assess the microcirculation, through direct visualization of the capillaries. Main patterns are classified, according to Cutolo et al., as scleroderma, further divided into early, active, or late patterns, or nonscleroderma. NFC findings include dilated loops, tortuosities, meandering or bushy capillaries, hemorrhage, or architectural distortion. NFC use has been indicated for the evaluation of Raynaud's phenomenon (RP), once it permits the distinction between primary and secondary RP. NFC results accounts for diagnostic criteria of systemic sclerosis, but they can also be useful in staging other connective tissue autoimmune diseases, like systemic lupus erythematosus, inflammatory myositis, or vasculitis. The CSURI index uses NFC for prediction of digital ulcer relapse. Recent evidence revealed NFC can also be applied in systemic disorders with vascular involvement.

Keywords: nailfold capillaroscopy, microcirculation, Raynaud's phenomenon, connective tissue autoimmune disorders, vasculitides

1. Introduction

Nailfold capillaroscopy (NFC) is a noninvasive, simple, and highly sensitive technique used in the study of microcirculation, as it permits direct visualization of nailfold capillaries [1, 2]. The understanding of NFC results from years of research in Raynaud's phenomenon (RP) in rheumatic diseases. In fact, this method is a paramount tool to differentiate between primary and secondary RP and, associated with autoantibodies, contributes for an early detection of systemic autoimmune connective tissue disorders (AICTD), as microcirculation abnormalities can arise as first manifestations of these diseases [3–5]. Its importance has reached a global recognition and validation as it became a classification criterion for systemic sclerosis, pointing 2 out of a minimum of 9 points to perform the diagnosis [6].

Ensuing studies have been disclosing the relationship between NFC abnormalities and some clinical syndromes or diseases, as digital ulcers, myositis, pulmonary hypertension, heart failure severity, diabetes mellitus, and arterial hypertension, among others [7–13]. Capillaroscopy can also be useful in monitoring the microvascular impact of certain drugs, as systemic vasodilators [2]. A role of NCF as a prognostic tool has been established with the Capillaroscopic Skin Ulcer Risk Index

(CSURI), to predict the appearance of new scleroderma ulcers and/or persistence of nonhealing lesions, within 3 months from NFC exam [14]. It has a good sensitivity, specificity, and positive predictive value, even in different devices. Its reliability has been successively demonstrated by European League Against Rheumatism (EULAR) study groups [15–17].

Indications for NFC do not resume to RP or other vascular acrosyndromes approach. It should be performed to any patient with microcirculation involvement from a systemic disease that includes AICTD, like systemic sclerosis, idiopathic inflammatory myositis, mixed connective tissue disease, and systemic lupus erythematosus, among others, but also other systemic diseases associated to microangiopathy, like vasculitides, diabetes, and arterial hypertension. As it also plays an important role in diagnosis, prognosis, and treatment monitoring of some diseases, capillaroscopy can be considered to act as a promising microcirculation biomarker [18].

2. Capillaroscopy procedures

NFC is commonly performed in nailfold cuticles, as the capillary loops become more parallel to the skin surface in this area and can be observed in their full length in the last row [2]. Usually, eight fingers are examined: the 2nd, 3rd, 4th, and 5th [1, 19]. The thumbs are excluded because, in these fingers, capillaries are poorly observed and microtrauma is more frequent due to thumb's opponency. At least four images should be taken from each finger, in order to maximize nailfold area visualization (**Figure 1**). Less than eight nailfold reduces the sensitivity to detect capillary changes [20]. So, in a regular exam, at least 32 pictures are taken from each hand when using a videocapillaroscope.

The patient should stay at least 20 minutes in a climatized room (20–22°C), to reduce RP attacks, and a clinical examination should be performed in his or her hands, in order to avoid traumatic injuries and to detect sclerodactyly, edema, pitting scars, active arthritis, skin lesions, onychophagia, or other possible changes [19, 21]. As several physiological and external factors can affect NFC image quality, patients are asked to avoid cutting cuticle or even the nails in the previous week, prevent nail varnish removal, and avoid smoking or drinking caffeine-containing beverages in the preceding hours [7]. To improve the amount of light reaching the nails and then ameliorate image quality, an immersing oil is used between the skin and the lens.

Different devices can be used for microcirculation visualization:

- Ophthalmoscopes and traditional microscopes: they consist in low cost options and are widely available, which can be used with minimal training; the disadvantages include not only low magnifications (10–20×), but also it is hard handling and it has a poor reproducibility.

- Dermatoscopes: they have an intermediate cost; they are portable devices and easily available and have acceptable resolution and sensitivity for NFC abnormalities; however, they also have magnification restraint (20–40×) and can only detect gross NFC changes, and images are not reproducible.

- Stereomicroscopes: they also have an intermediate cost, they are easy to use, their magnification lens vary from 10–200×, but they are time-consuming and difficult to use in patients with joint contractures and they need an additional camera and fiber optic light source to capture the images, and a specialized training is even required.

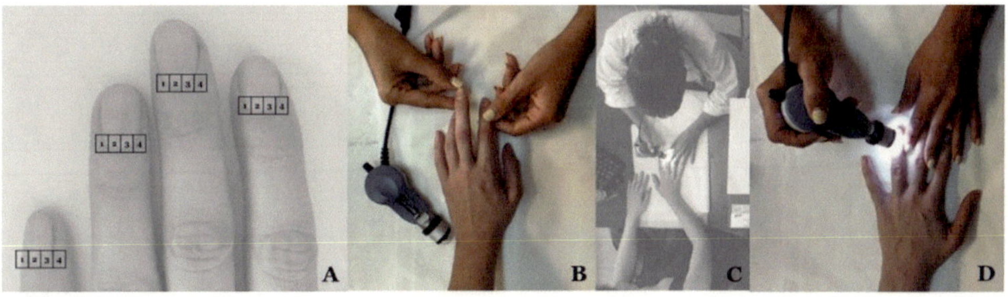

Figure 1.
Nailfold capillaroscopy should be performed in 2nd, 3rd, 4th, and 5th fingers, dividing each nailfold cuticle into 4 areas and taking images from each of them. (A) Clinical examination of the fingers should be performed before the exam begins. (B) Performing videocapillaroscopy using Videocap biomicroscope, version 3.0. (C, D).

- Videocapillaroscope (**Figure 1**): it has an excellent image quality and reproducibility and variable magnification from 200–600×. It has a portable system enabling its use in patients with severe joint contractures, and it carries specific software for capturing and analyzing the images, yet it is the most expensive option and requires specialized training [7].

NFC procedure includes the assessment of morphologic and some functional parameters: skin transparency, capillary density and orientation, venous plexus visualization, presence of neoangiogenesis or microhemorrhages, capillary loop diameter, length and morphology, and characteristics of blood flow [1, 21]. The number of capillaries has been validated as the most important criterion for the patient follow-up [22].

3. Nailfold capillaroscopy assessment and classification

NFC studies lead to the detection of microvascular abnormalities. Some of them, rarer, have a clear pathological significance and can disclose early an AICTD-they are called the "major abnormalities" [1]. Among those are giant capillaries, capillary architecture disorganization, microhemorrhages, neoangiogenesis, and capillary loss. Other abnormalities, more frequent, have an uncertain pathological meaning and represent an overlap between the scope of normality and microangiopathy-they are called the "minor abnormalities." Those are principally tortuosity, abnormal shapes, or visibility of the subpapillary venous plexus.

Capillaroscopic parameters are usually evaluated through qualitative, semiquantitative, and quantitative analysis. The qualitative assessment implies pattern recognition and describes the global microvascular array and architecture, shape and distribution of the capillaries, and abnormalities of single loops. It readily allows a distinction between a normal capillaroscopy exam and an abnormal one [20]. The semiquantitative analysis is focused on major NFC changes. The quantitative evaluation estimates capillary density, avascular areas, diameter of enlarged capillaries, and the frequency of each abnormality. The quantification of capillaries includes the number of loops in the distal row in a 1 mm^2 section of the nailfold (**Figure 2**).

Normal capillaries are hairpin shaped and present a homogeneous distribution, in a "comb-like" structure [1] (**Figure 3**). Capillary density varies from 9 to 14 capillaries per millimeter in adults and at least 6 in children [1, 7, 23]. Abnormal shapes include the following:

- **Tortuous capillaries**: arterial and venous limbs are curled but do not cross [23] (**Figure 4A**).

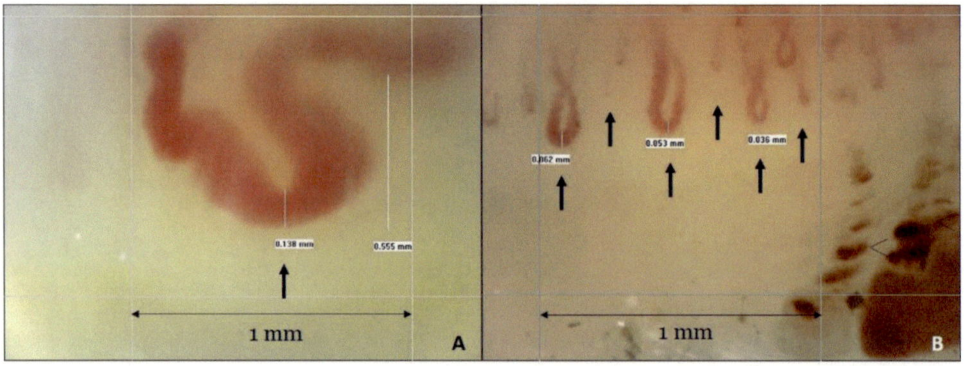

Figure 2.
Capillary density. Panel A: 1 capillary/mm²; panel B: 6 capillaries/mm². (arrows) images were taken using Videocap biomicroscope, version 3.0, magnification ×200. Courtesy of Nailfold Capillaroscopy Clinic of Hospital Curry Cabral.

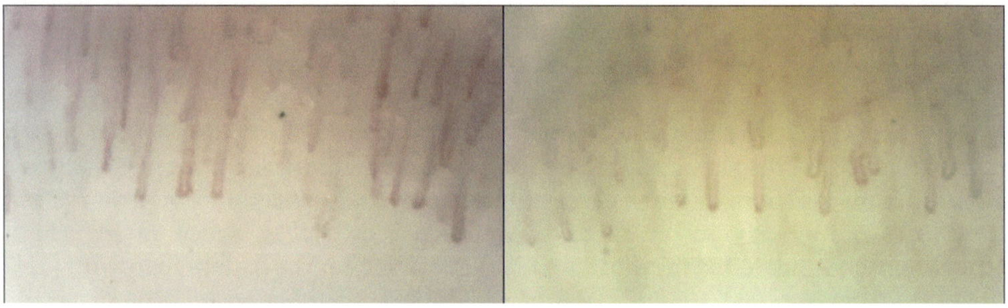

Figure 3.
Examples of images of nailfold capillaroscopy captured in healthy subjects and considered as normal. Images were taken using Videocap biomicroscope, version 3.0, magnification ×200. Courtesy of Nailfold Capillaroscopy Clinic of Hospital Curry Cabral.

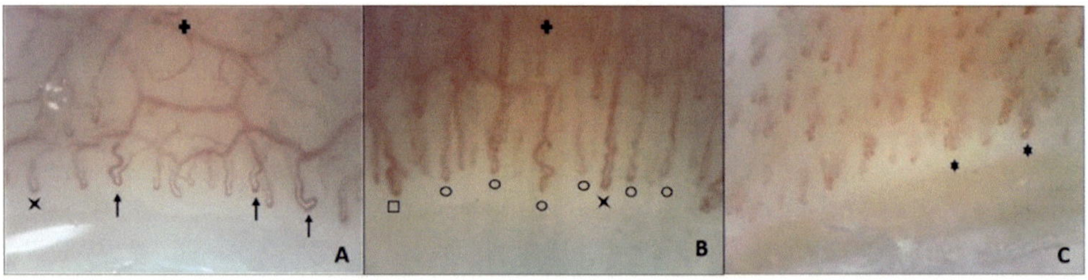

Figure 4.
Examples of capillaroscopy abnormalities: Tortuosity (arrow), crossing (X), meandering (black circle), bushy capillaries (six-pointed star), subpapillary plexus visibility (plus sign), and dilated loop (black square). Images were taken using Videocap biomicroscope, version 3.0, magnification ×200. Courtesy of Nailfold Capillaroscopy Clinic of Hospital Curry Cabral.

- Crossing capillaries: arterial and venous limbs cross at one point [23] (**Figure 4B**).

- **Meandering loops**: the limbs cross upon themselves or with each other several times [23] (**Figure 4A**).

- Bushy capillaries: the limb branches present themselves in small and multiple buds [7] (**Figure 4C**).

- **Ramified capillaries:** there are abnormal connections between arterial and venous limbs or different capillaries [7] (**Figure 5**).
- **Neoangiogenesis:** neoformations can be heterogeneous and may comprise shape heterogeneity, four or more capillaries within a single dermal papilla, elongated loops, branching, and interconnected capillaries [1] (**Figure 5**).
- **Bizarre loops**: capillaries with an atypical morphology, distinct from other described categories [23].
- **Dilated or enlarged loops**: there is not a universally accepted definition for loop enlargement, but it is usually considered if the limbs are enlarged about four times the normal width or if the width diameter is >20 μm. Nevertheless, a recent study from Cutolo et al suggests that dilations below 30 μm may be considered nonspecific [23] (**Figure 6**).
- **Megacapillary or giant capillary**: capillaries with homogeneously enlarged loops with a branch diameter above 50 μm [1, 23] (**Figure 6**).
- Elongated capillaries: capillary loops longer than 300 μm [7] (**Figure 7**).

Other capillaroscopic parameters include the following:

- **Hemorrhages**: microbleeding appears as dark masses adjacent to distal row, due to hemosiderin deposits. They result from disruption of the capillary wall, either spontaneous or traumatic [1, 7] (**Figure 8**).
- **Avascular areas**: lack of two or more successive capillaries. Loss of capillaries is associated with tissue hypoxia and subsequent digital ulcers and ischemia [7, 23] (**Figure 2**).
- **Subpapillary venous plexus visibility**: observation of large and linked arrangement of vessels with a greater caliber than the capillaries. Enlargement and congestion of venules and capillaries related to persistent opening of arteriovenous anastomoses, thus enabling a greater visibility [1] (**Figure 4**).
- **Capillary blood flow**: with powerful magnifications, when the blood flow is slow, it is possible to see the packs of red blood cells moving as a capillary sludge [1] (**Figure 9**).
- **Capillary array and architecture disarrangement**: when shapes, length, and diameter vary in continuous loops, it leads to a complete distortion of a normal capillary pattern [1] (**Figure 10**).

Major NFC patterns divide into scleroderma and nonscleroderma pattern. Scleroderma pattern was first described by Maricq et al, through a combination of widening of the capillary loop, loss of capillaries, and disorganization of the nailfold capillary bed [1]. This pattern is frequently seen in scleroderma spectrum disorders, like systemic sclerosis, dermatomyositis, and mixed connective tissue disease and in RP without a clear diagnosis. Later, Cutolo et al further classified the scleroderma pattern into "early," "active," and "late" stages. "Early" scleroderma pattern presents with few giant capillaries and hemorrhages, relatively well-preserved capillary

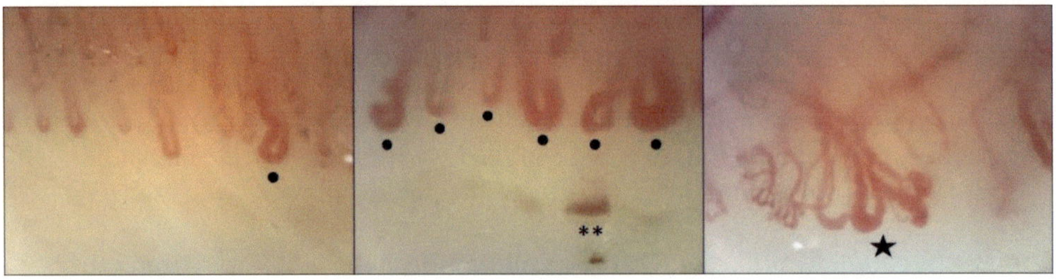

Figure 5.
Images from three different phases of scleroderma pattern, belonging to three different patients: Early, active, and late (from left to right). Capillary density decreases from early to late pattern. Capillaroscopic findings include giant capillaries (black spot), hemorrhages (**), and neoangiogenesis (star). Images were taken using Videocap biomicroscope, version 3.0, magnification ×200. Courtesy of Nailfold Capillaroscopy Clinic of Hospital Curry Cabral.

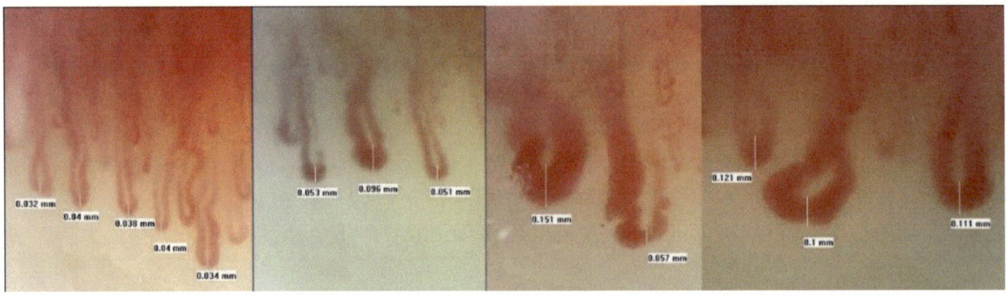

Figure 6.
Capillary limb dilatation: Dilated (diameter ≥ 30 μm) and giant (diameter ≥ 50 μm) capillaries. Images were taken using Videocap biomicroscope, version 3.0, magnification ×200. Courtesy of Nailfold Capillaroscopy Clinic of Hospital Curry Cabral.

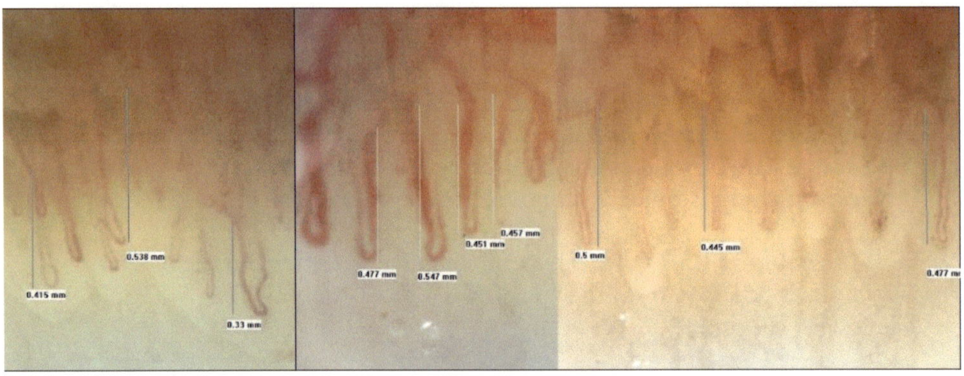

Figure 7.
Elongated capillaries (capillary loops ≥300 μm). Images were taken using Videocap biomicroscope, version 3.0, magnification ×200. Courtesy of Nailfold Capillaroscopy Clinic of Hospital Curry Cabral.

distribution, and no loss of capillaries. "Active" scleroderma pattern reveals frequent giant capillaries and hemorrhages, moderate loss of capillaries, mild disorganization of capillary bed, and absent or mild ramified capillaries (i.e., neoangiogenesis). "Late" scleroderma pattern shows an irregular enlargement of the capillaries, few or absent giant capillaries and hemorrhages, severe loss of capillaries with avascular areas (the *plages désertes*), and ramified or bushy capillaries (**Figure 5**).

In healthy subjects, anatomic variations occur, and capillary abnormalities can be found. The estimated prevalence of these changes are meandering loops in 25%,

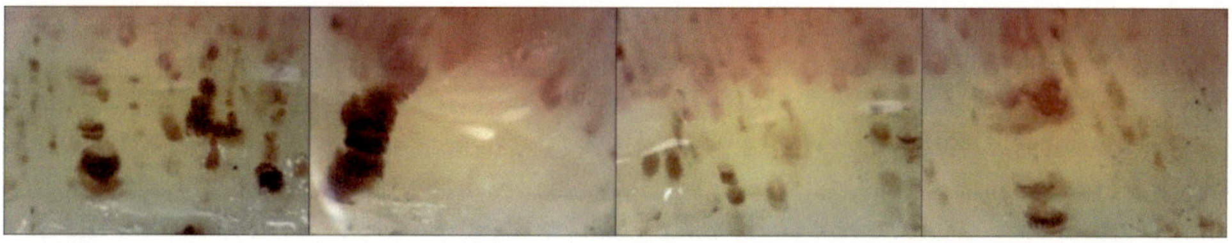

Figure 8.
Examples of nailfold hemorrhages in patients with different diseases. Images were taken using Videocap biomicroscope, version 3.0, magnification ×200. Courtesy of Nailfold Capillaroscopy Clinic of Hospital Curry Cabral.

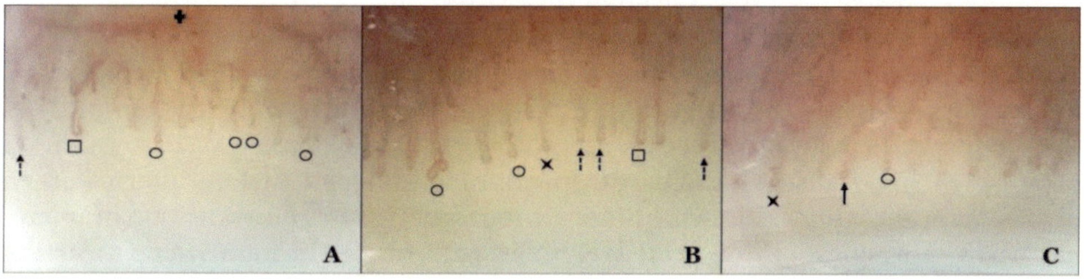

Figure 9.
Images from a single patient with Sjögren syndrome, presenting dilated loops (black square), meandering (black circle), crossing (X), reduced capillary flow with sludge (dashed arrow), and ingurgitated venous plexus (plus sign). (A: 5th finger of the right hand; B: 2nd finger of the left hand; C: 4th finger of the right hand.) images were taken using Videocap biomicroscope, version 3.0, magnification ×200. Courtesy of Nailfold Capillaroscopy Clinic of Hospital Curry Cabral.

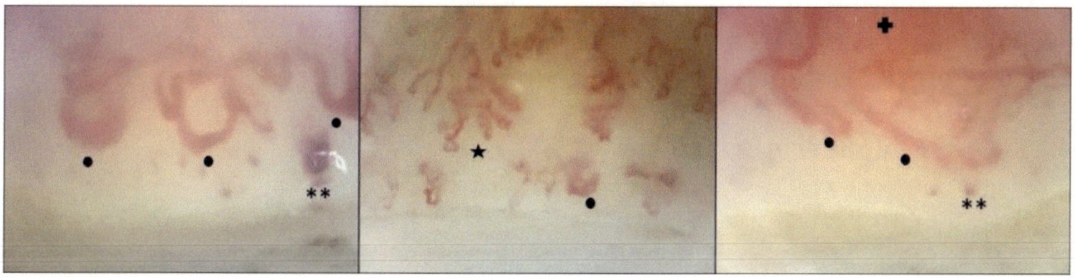

Figure 10.
Dermatomyositis-images from 3 different patients, revealing reduced capillary density, with a complete distortion of the vascular array, neoangiogenesis (star), giant capillaries (black dot), and hemorrhages (**). Venous plexus is also ingurgitated (plus sign). Images were taken using Videocap biomicroscope, version 3.0, magnification ×200. Courtesy of Nailfold Capillaroscopy Clinic of Hospital Curry Cabral.

dilated capillaries in 12%, bushy loops in 7%, bizarre loops in 2%, and giant capillaries in 0.3% of the healthy population [23].

4. Nailfold capillaroscopy in autoimmune connective tissue diseases

AICTD complex pathogenesis usually includes microvascular changes, with occurrence of progressive structural and functional damage of the capillaries. Therefore, NFC became an important diagnostic and prognostic tool to use while managing these disorders.

4.1 Systemic sclerosis (SSc)

Systemic sclerosis has probably been the most studied disease with NFC. It is a severe AICTD in which the main pathological events are endothelial dysfunction, fibrosis, and inflammation, which usually results in skin and vascular changes [24]. These include an important damage of microvascular network, with enlarged and giant capillaries, capillary loss with disarrangement of capillaries' architecture, and neoangiogenesis. As mentioned above, the scleroderma patterns are divided into three different patterns: "early," revealing only few enlarged capillaries, few hemorrhages, and preserved capillary density; "active," showing frequent giant capillaries and hemorrhages, mild capillary loss, and disorganization of the microvascular network; and "late," where irregular and giant capillaries can be found, as with few hemorrhages, neoangiogenesis, and avascular areas [25] (**Figure 5**).

A large multinational study based on EULAR Scleroderma Trials and Research (EUSTAR) registry disclosed a scleroderma pattern in more than 86% of SSc patients [24]. Subjects without this pattern did not have organ involvement and RP and some had negative autoantibodies. These results suggest that although these patients were classified as having an SSc, they did not have an overt disease and, then, nonspecific NFC changes may precede an evolving scleroderma pattern. In patients with overt disease, capillaroscopic findings mirror somehow internal organ involvement evolution. Following disease progression, dynamic transition of microvascular abnormalities through different NFC patterns can be found in up to 50% of SSc patients [18, 26]. On the other hand, capillaroscopy patterns can improve after up to 4 years of combined treatment, revealing a progressive significant recovery in structure and function of microcirculation, associated to ameliorated outcomes, independently of disease severity [27–29].

Capillaroscopy abnormalities became one of the diagnostic criteria for SSc in the 2013 classification [6]. NFC should not only be used for a diagnostic purpose but also for monitoring the disease process and determining its prognosis, because, as explained above, its dynamic changes occur, and its severity directly relates to the extent of organ involvement [8].

4.2 Systemic lupus erythematosus (SLE)

A systematic review was recently performed by a EULAR study group, in order to establish capillaroscopic parameters in SLE patients and its correlations with clinical and laboratory characteristics [30]. According to this study, SLE patients present more tortuous and abnormal capillaries than healthy controls, as well as more hemorrhages (**Figure 11**). An NFC score was created by these authors to set the microangiopathy severity, SLE patients being those who had the higher scores. "Nonspecific patterns" and "scleroderma-like patterns" were also described. A correlation between NFC abnormalities and clinical and laboratory parameters was established, since a relationship between NFC score and SLE activity was disclosed. Further, it is an important note to highlight that, once SLE is a heterogeneous disease, with altered vascular involvement, probably capillaroscopic changes will only be seen in the active phase of the disease [31]. Also, as the microvasculopathy profile in SLE is quite different from the SSc's, which is typically obliterative, changes as neoangiogenesis are less common in SLE patients [32, 33].

4.3 Inflammatory idiopathic myositis

In inflammatory idiopathic myopathies (IIM), it is frequent to find tortuosities, capillary loss, enlarged and giant capillaries, microhemorrhages, and neoangiogenesis, as well as a disorganization of the vascular network and avascular areas [1, 34]. In dermatomyositis (DM), patients present more severe NFC findings, compared with those with polymyositis (PM). Ramified and bushy capillaries represent a hallmark of microvascular damage in DM (**Figure 10**). In these patients, capillaroscopic abnormalities seem to be related with disease duration: in the first 6 months of disease duration, capillary density is usually reduced and giant capillaries are frequent; after that period, scleroderma pattern becomes more common. In PM patients, NFC findings do not significantly differ from healthy controls. It has also been demonstrated that there is a potential relationship between capillary changes and organ involvement, especially in patients with lung disease [35].

A recent multicenter study in antisynthetase syndrome revealed that NFC changes are usually independent from the presence of RP [27]. In these patients, the scleroderma pattern is associated to positivity for anti-Jo1 antibodies and a longer disease duration. An interesting finding was that significant correlation was established between ILD and ramified capillaries, but not with SSc-like pattern.

Together, these studies suggest that NFC can become an important indicator of interstitial lung disease in patients with IIM, disclosing early this potential life-threatening manifestation.

4.4 Mixed connective tissue disease (MCTD)

In MCTD, different abnormalities can be found: minor changes, hemorrhages, dilated and giant capillaries, reduced density, and neoangiogenesis [19] (**Figure 12**). The patients can present with a scleroderma-like pattern, but less severe when compared to systemic sclerosis. The dilated loops are usually more dystrophic, and the dilated loops are long. The neoangiogenesis features are normally present in patients who progress to SSc. The avascular areas are more frequent in patients with pulmonary involvement or under immunosuppressants [36].

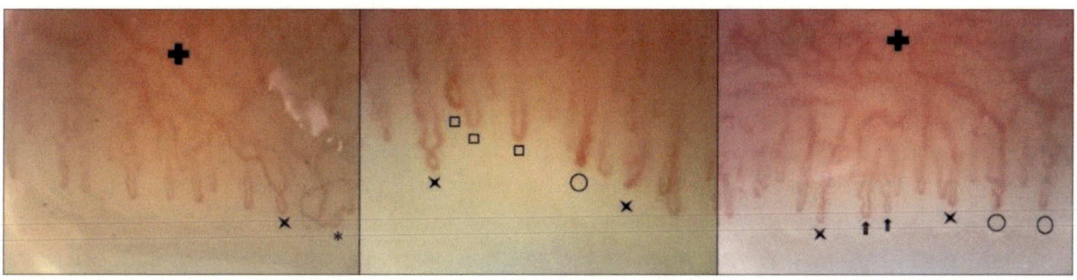

Figure 11.
Systemic lupus erythematosus. Images from three different patients, revealing crossing capillaries (x), dilated loops (square), meandering capillaries (black circle), tortuosities (arrow), ingurgitated venous plexus (plus sign), and loop aneurysm (*). Images were taken using Videocap biomicroscope, version 3.0, magnification ×200. Courtesy of Nailfold Capillaroscopy Clinic of Hospital Curry Cabral.

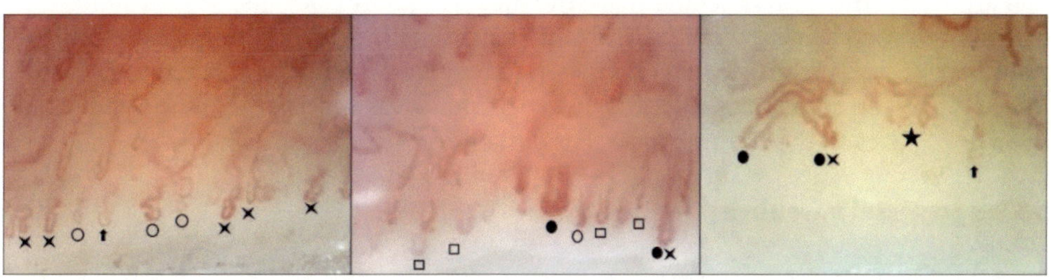

Figure 12.
Images from 3 different patients with mixed connective tissue disease. Capillaroscopic findings include crossing (x), meandering (black circle), tortuosity (arrow), dilated loops (black square), giant capillaries (black dot), neoangiogenesis (black star), and clear avascular areas on the right image. Images were taken using Videocap biomicroscope, version 3.0, magnification ×200. Courtesy of Nailfold Capillaroscopy Clinic of Hospital Curry Cabral.

4.5 Sjögren syndrome

In patients with Sjögren syndrome (SSj), NFC can be normal in up to 59% of cases, if RP is absent [19]. The more frequent findings in this disease are tortuosities, crossings, and ingurgitated venous plexus (**Figure 9**). In patients carrying positive anticentromere antibody, dilated loops and giant capillaries can also be found. No correlations were stated between NFC changes and laboratory parameters.

4.6 Rheumatoid arthritis

The different abnormalities found in rheumatoid arthritis patients confirm the coexistence of microangiopathy in this disease [37]. In patients without rheumatoid vasculitis, it is frequent to find thin, long, and tortuous capillaries, with ingurgitated anarchic venous plexus and microhemorrhages [19]. These changes have no correlation with disease activity. Dilated or giant capillaries are rare and justify a closer follow-up.

4.7 Psoriasis

NFC in psoriasis usually reveals a reduced capillary density, with avascular areas, and morphologically abnormal capillaries [38]. No correlation was found between capillary density and disease duration or the extent of skin involvement, but avascular areas are more frequent in patients whose nails are also affected. If the exam is performed over the psoriasis plaques, dilated and long loops can be seen, with interstitial edema and fast blood flow [19].

5. Nailfold capillaroscopy in noninfectious vasculitides

Vasculitides evolve by inflammation of vessels, which include capillaries. NFC can, then, provide valuable information on the approach of patients with vasculitis. Although scarce research has been made in this field, a recent systematic review, following Chapel-Hill nomenclature, puts in evidence that NFC is more useful in small than in large vessel vasculitides [39, 40]. However, as large vessel vasculitides also involve microcirculation, the presence of NFC changes in these disorders cannot be excluded. Besides, NFC can also give important information about organ involvement and disease activity [41, 42].

The microangiopathy in vasculitides reveals several and heterogeneous NFC changes, but generally nonspecific, when compared to scleroderma pattern. Increased tortuosity; microhemorrhages; enlarged, bushy, and bizarre capillaries; and architectonic disarrangement are the most frequent findings [39]. In some cases, however, the described NFC abnormalities include capillary dilation and reduced density, which are usually associated with scleroderma spectrum diseases. Whether a deep capillary damage is due to primary vasculitis or if there is a possible overlap of cases with scleroderma-related disorders is yet to be clarified.

5.1 Large vessel vasculitis

A recent study revealed that in Takayasu arteritis (TA) capillaries are affected due to hypoperfusion [43]. NFC abnormalities found were reduced capillary length and venous limb diameter and tortuosity, but, in hands with subclavian involvement, these changes were more severe. Capillary diameter was then considered as an example of subclavian artery stenosis alteration due to disease progression.

We found no data about nailfold capillaroscopy in giant cell arteritis.

5.2 Medium vessel vasculitis

In polyarteritis nodosa, NFC can be normal in the absence of Raynaud's phenomenon [19]. Yet, in its presence, changes include reduced capillary density, microhemorrhages, and edema. When digital ischemia is present, important edema, capillary flow sludge, and multiple hemorrhages can be seen (**Figure 13**).

Only one study about Kawasaki disease and NFC was found and it involved 64 pediatric patients [44]. Microcirculation abnormalities found included reduced density, dilation of arterial and venous limb diameters, higher intercapillary distance, and abnormal loops. The latter two were related to disease activity, as they improved from postacute to convalescent phase. Blood velocity was associated with increased coronary artery diameter.

5.3 Small vessel vasculitis

- Antineutrophil cytoplasmic antibody (ANCA)-associated vasculitis

Microcirculation abnormalities of granulomatosis with polyangiitis, formerly Wegener's granulomatosis, were described in one study involving 12 patients [45]. The main NFC changes detected were avascular areas; crossed and bushy capillaries; and microhemorrhages. No relationship was established with disease activity or its clinical aspects.

No valuable information was found about NFC in microscopic polyangiitis and eosinophilic granulomatosis with polyangiitis-former Churg-Strauss vasculitis.

- Immune complex small vessel vasculitis

NFC changes in cryoglobulinemic vasculitis were disclosed in one study with 29 patients, of which 28 had hepatitis C infection [41]. Microcirculation abnormalities detected were tortuosity, altered orientation, shortened capillaries, and neoangiogenesis (**Figure 14**). No relation was found with disease activity, but glomerulonephritis was associated with a higher score of NFC alterations.

IgA vasculitis (IAV), formerly called Henoch-Schönlein purpura, rarely affects adults and studies about NFC changes have been performed in small samples of patients in pediatric age [46–48]. The NFC changes in IAV are conflicting, including density reduction, increased capillary length, loop dilatation, persistent edema,

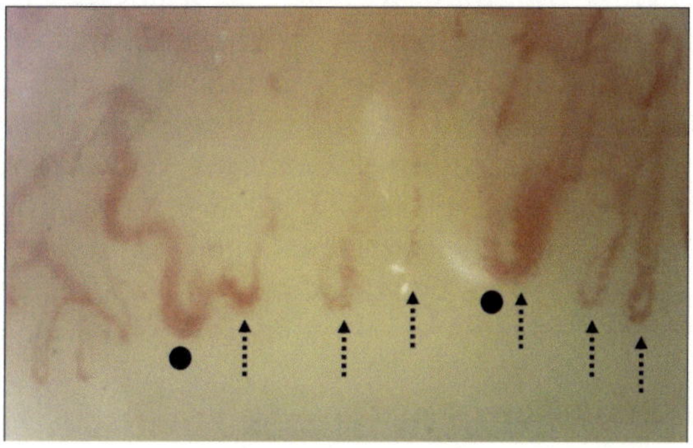

Figure 13.
Polyarteritis nodosa. This patient presents with capillary slow flow sludge (dashed arrow), reduced capillary density, but also enlarged capillaries (black dot) (4th finger of the left hand). Images were taken using Videocap biomicroscope, version 3.0, magnification ×200. Courtesy of Nailfold Capillaroscopy Clinic of Hospital Curry Cabral.

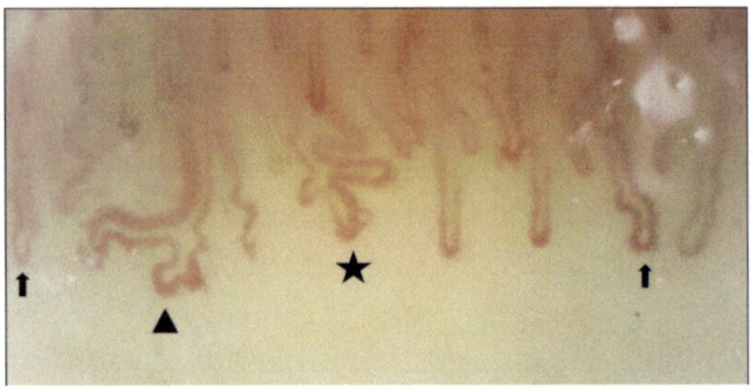

Figure 14.
Cryoglobulinemic vasculitis. The image reveals tortuosity (arrow), bifurcation with altered orientation (triangle), and neoangiogenesis (star) (2nd finger of the right hand). Images were taken using Videocap biomicroscope, version 3.0, magnification ×200. Courtesy of Nailfold Capillaroscopy Clinic of Hospital Curry Cabral.

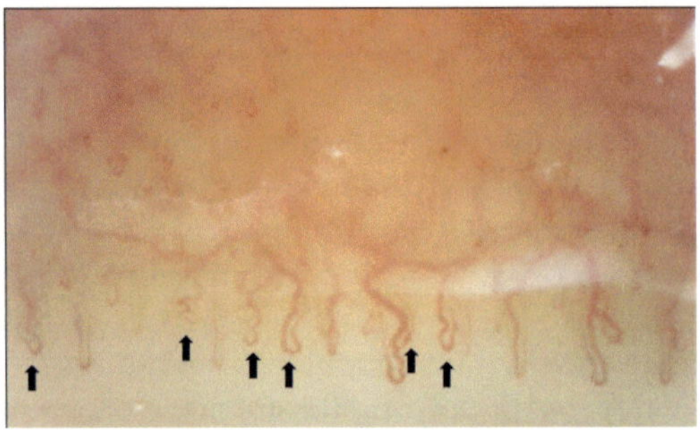

Figure 15.
Tortuosity (arrow) in a patient with Behçet disease (4th finger of the left hand). Images were taken using Videocap biomicroscope, version 3.0, magnification ×200. Courtesy of Nailfold Capillaroscopy Clinic of Hospital Curry Cabral.

tortuosity, and branching capillaries. No statistically significant correlation was found between NFC abnormalities and organ involvement, but Zampetti et al. described normalization of the edema after 6-month follow-up.

There was no available data about NFC in anti-glomerular basement membrane disease and hypocomplementemic urticarial vasculitis-former anti-C1q vasculitis.

5.4 Variable vessel vasculitis

Behçet disease (BD) relationship with NFC alterations have been described in some studies, but they all used different technical characteristics for visualization of microcirculation [42, 49–51]. The main NFC abnormalities include enlarged capillaries, microhemorrhages, reduced density, and tortuosity (**Figure 15**). None of them were related with disease activity. Still, some NFC alterations were described as being related with clinical aspects of BD: NFC severity corresponded to longer disease duration and positive pathergy test; enlarged capillaries were associated with younger age at disease onset, high blood pressure, and superficial phlebitis.

We did not find any information about capillaroscopy in Cogan syndrome.

5.5 Vasculitis associated with systemic disease

In rheumatoid vasculitis, some studies correlated with capillary damage and levels of soluble intracellular adhesion molecule-1 (sICAM-1), which is highly expressed during inflammation [52–54]. They found abnormalities in the great majority of patients, mainly morphologic changes, and although there was no direct relation with disease activity, severe NFC alterations were associated with disease duration, cutaneous vasculitis, joint erosions, systemic vasculitis, and sICAM-1 levels. In a recent study involving 62 patients, scleroderma-like NFC changes were found in 20% of patients with rheumatoid vasculitis and they were interpreted as varying degree of microvascular inhomogeneity, not being necessarily related to overlap syndromes [55, 56].

No specific information was found about NFC in lupus vasculitis or sarcoid vasculitis.

6. Nailfold capillaroscopy in systemic disorders with vascular involvement

6.1 Antiphospholipid syndrome (APS)

Since APS does not derive from connective tissue, we consider more appropriated to approach it in a separated part of this chapter. Capillaroscopy has been studied in APS and attempts were made to include it as a diagnostic tool [57]. NFC findings include microhemorrhages and dilated loops. Long loops and slow flow sludge capillaries are suggestive of a primary APS, while hemorrhages are typical of secondary APS [19] (**Figure 16**). A specific pattern of microhemorrhage, symmetrically disposed, has been called the "comb-like" hemorrhage and is highly associated to APS [1]. Further, positivity for anticardiolipin antibody has been related to higher prevalence of hemorrhages [58]. In spite of this, and even if microhemorrhages significantly correlate with the diagnosis of APS and its clinical manifestations, NFC findings are not sufficient to establish APS diagnosis for its lack of sensitivity and specificity.

6.2 Diabetes mellitus (DM)

In DM, NFC changes are apparently associated to the level of glycemic control and the existence of chronic microvascular complications [9]. However, there is an elevated prevalence of comorbidities concurring for microangiopathy, especially in type 2 diabetes, including arterial hypertension, dyslipidemia, and obesity. Still, a "diabetic capillaropathy" was described, which includes tortuosity, capillaries with bizarre shapes, loop dilations, and avascular areas [9, 59]. No differences were found between type 1 and 2 DM, but microvascular complications detected with NFC were correlated with diabetic peripheral neuropathy [60]. It has also been demonstrated that even in prediabetic patients, microangiopathy can already be

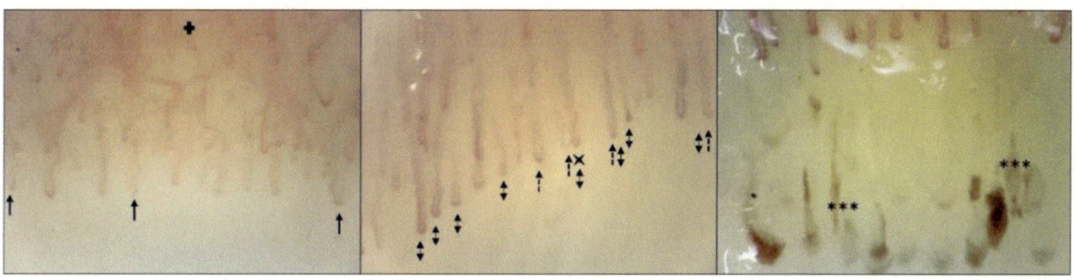

Figure 16.
Antiphospholipid syndrome in three different patients, revealing tortuosity (arrow), ingurgitated venous plexus (plus sign), long capillaries (two-way arrow), slow flow sludge (dashed arrow), and "comb-like" hemorrhages (***). Images were taken using Videocap biomicroscope, version 3.0, magnification ×200. Courtesy of Nailfold Capillaroscopy Clinic of Hospital Curry Cabral.

detected. So, it is reasonable to suggest that capillaroscopy could be included in the screening of DM-related complications, since it can easily detect microvascular damage at a peripheric level.

6.3 Arterial hypertension

The available data on NFC in arterial hypertension is scanty, but microangiopathy has been demonstrated in different studies. Decreased capillary density has been described as being related with the mean diastolic pressure and the blood hyperviscosity [61]. These abnormalities are even more severe in the hypertensive elderly population, once capillary loops become longer and narrower. A slow flow capillary sludge is usually found in individuals with a hypertensive systolic pressure. Capillary density is reversible, as it has been proved in different ethnic groups that a reduction in ingested salt would result in augmented capillary density [62]. A different study revealed a significant narrowing of arterial loops in patients with either arterial hypertension or prehypertension stage, regardless of their age [11].

7. Conclusion

Capillaroscopy is a paramount tool in the microcirculation study. Its reliability in the early diagnosis of some AICTD disorders has already been well stated, but it has also proved its great value in treatment monitoring and predicting systemic complications, especially in scleroderma spectrum disorders. In this way, it should be considered as a potential biomarker for microangiopathy.

Until now, scarce data are available about the impact of capillaroscopy on vasculitides approach. However, some studies have shown an active involvement of capillaries in these disorders, apart from the size of the vessels typically involved, which offers a large field and opportunity for further investigation. Large-scale standardized studies are, thus, required to clarify the role of capillaroscopy in vasculitis.

Conflict of interest

The authors declare no conflict of interest.

Author details

Vera Bernardino*, Ana Rodrigues, Ana Lladó, Melissa Fernandes and António Panarra
Internal Medicine 7.2 Department, Nailfold Capillaroscopy Clinic, Hospital Curry Cabral – Centro Hospitalar Universitário Lisboa Central, Lisbon, Portugal

*Address all correspondence to: verarodriguesb@gmail.com

References

[1] Cutolo M. Atlas of Capillaroscopy in Rheumatic Diseases. Milano: Elsevier; 2012

[2] Tavakol ME, Fatemi A, Karbalaie A, et al. Nailfold Capillaroscopy in rheumatic diseases: Which parameters should be evaluated? Journal of Biomedicine and Biotechnology. 2015;**2015**:974530

[3] Koenig M, Joyal F, Fritzler MJ, et al. Autoantibodies and microvascular damage are independent predictive factors for the progression of Raynaud's phenomenon to systemic sclerosis: A twenty-year prospective study of 586 patients, with validation of proposed criteria for early systemic sclerosis. Arthritis and Rheumatism. 2008;**58**(12):3902-3912

[4] Cutolo M, Pizzorni C, Tuccio M, et al. Nailfold videocapillaroscopy patterns and serum autoantibodies in systemic sclerosis. Rheumatology. 2004;**43**(6):719-726

[5] Ingegnoli F, Borachi P, Gualtierotti R, et al. Improving outcome prediction of systemic sclerosis from isolated Raynaud's phenomenon: Role of autoantibodies and nailfold capillaroscopy. Rheumatology. 2010;**49**(4):797-805

[6] Van den Hoogen F, Khanna D, Fransen J, et al. 2013 classification criteria for systemic sclerosis: An American college of rheumatology/ European league against rheumatism collaborative initiative. Arthritis and Rheumatism. 2013;**65**(11):2737-2747

[7] Karbalaie A, Emrani Z, Fatemi A, et al. Practical issues in assessing nailfold capillaroscopic images: A summary. Clinical Rheumatology. 2019;**38**:2343-2354. DOI: 10.1007/s10067-019-04644-9

[8] Soulaidopoulos S, Triantafyllidou E, Garyfallos A, et al. The role of Nailfold capillaroscopy in the assessment of internal organ involvement in systemic sclerosis: A critical review. Autoimmunity Reviews. 2017;**16**:787-795

[9] Lisco G, Cicco G, Cignarelli A, et al. Computerized video-capillaroscopy alterations related to diabetes mellitus and its complications. Advances in Experimental Medicine and Biology. 2018;**1072**:363-368

[10] Emrani Z, Karbalaie A, Fatemi A, et al. Capillary density: An important parameter in nailfold capillaroscopy. Microvascular Research. 2017;**109**:7-18

[11] Gurfinkel Y, Sasonko ML, Priezzhev AV. Digital capillaroscopy as important tool for early diagnostics of arterial hypertension. Proceedings of SPIE. 2015;**9448**:944804

[12] Shirshin EA, Gurfinkel YI, Matskeplishvili ST, et al. In vivo optical imaging of the viable epidermis around the nailfold capillaries for the assessment of heart failure severity in humans. Journal of Biophotonics. 2018;**11**(9):e201800066

[13] Ghizzoni C, Sebastiani M, Manfredi A, et al. Prevalence and evolution of scleroderma pattern in nailfold videocapillaroscopy in systemic sclerosis patients: Clinical and prognostic implications. Microvascular Research. 2015;**99**:92-95

[14] Sebastiani M, Manfredi M, Lo Monaco A, et al. Capillaroscopic skin ulcers risk index (CSURI) calculated with different videocapillaroscopy devices: How its predictive value change. Clinical and Experimental Rheumatology. 2013;**31**(2;76):115-117

[15] Cutolo M, Melsens K, Herrick A, et al. Reliability of simple capillaroscopic definitions in describing capillary morphology in rheumatic

diseases. Rheumatology. 2018;**57**(4):757-759

[16] Smith V, Beeckman S, Herrick A, et al. An EULAR study group pilot study on reliability of simple capillaroscopic definitions to describe capillary morphology in rheumatic diseases. Rheumatology. 2016;**55**(5):883-890

[17] Cutolo M, Smith V. State of art on nailfold capillaroscopy: A reliable diagnostic tool and putative biomarker in rheumatology? Rheumatology. 2013;**52**(11):1933-1940

[18] Smith V, Decuman S, Sulli A, et al. Do worsening scleroderma capillaroscopic patterns predict future severe organ involvement? A pilot study. Annals of the Rheumatic Diseases. 2012;**71**:1636-1639

[19] Jammal M, Kettaneh A, Cabane J, et al. Capillaroscopie périunguéale: une évaluation simple et fiable de toute pathologie de la microcirculation. La Revue de Médecine Interne. 2015;**36**:603-612

[20] Cutolo M, Sulli A, Smith V, et al. How to perform and interpret capillaroscopy. Best Practice & Research. Clinical Rheumatology. 2013;**27**(2):237-248

[21] FAggioli P, Tamburello A, Sciascera A, et al. Nailfold videocapillaroscopy in internal medicine. Italian Journal of Medicine. 2015;**9**:234-242

[22] Cutolo M, Herrick AL, Distler O, et al. Nailfold videocapillaroscopic features and other clinical risk factors for digital ulcers in systemic sclerosis: A multicenter, prospective cohort study. Arthritis & Rheumatology. 2016;**68**(10):2527-2539

[23] Cutolo M, Sulli A and Smith V. Capillary microscopy. In Atlas of Ulcers in Systemic Sclerosis, Springer Nature Switzerland AG 2019, M. Matucci-Cerinic, CP Denton

[24] Ingegnoli F, Ardoino I, Boracchi P. Nailfold capillaroscopy in systemic sclerosis: Data from the EULAR scleroderma trials and research (EUSTAR) database. Microvascular Research. 2013;**89**:122-128

[25] Cutolo M, Sulli A, Smith V. Assessing microvascular changes in systemic sclerosis diagnosis and management. Nature Reviews Rheumatology. 2010;**6**:578-587

[26] Sulli A, Pizzorni C, Smith V, et al. Timing of transition between capillaroscopic patterns in systemic sclerosis. Arthritis and Rheumatism. 2012;**64**(3):821-825

[27] Cutolo M, Zampogna G, Vremis L, et al. Longterm effects of endothelin receptor antagonism on microvascular damage evaluated by nailfold capillaroscopic analysis in systemic sclerosis. The Journal of Rheumatology. 2013;**40**:40-45

[28] Cutolo M, Ruaro B, Pizzorni C, et al. Longterm treatment with endothelin receptor antagonist bosentan and iloprost improves fingertip blood perfusion in systemic sclerosis. The Journal of Rheumatology. 2014;**41**:881-886

[29] Trombetta AC, Pizzorni C, Ruaro B, et al. Effects of longterm treatment with bosentan and iloprost on nailfold absolute capillary number, fingertip blood perfusion, and clinical status in systemic sclerosis. The Journal of Rheumatology. 2016;**43**:2033-2041

[30] Cutolo M, Melsens K, Wijnant S, et al. Nailfold capillaroscopy in systemic lupus erythematosus: A systematic review and critical appraisal. Autoimmunity Reviews. 2018;**17**:344-352

[31] Soltesz P, Kerekes G, Der H, et al. Comparative assessment of vascular function in autoimmune rheumatic diseases: Considerations of prevention and treatment. Autoimmunity Reviews. 2011;**10**:416-425

[32] Murdaca G, Colombo BM, Cagnati P, et al. Endothelial dysfunction in rheumatic autoimmune diseases. Atherosclerosis. 2012;**224**:309-317

[33] Chora I, Guiducci S, Manetti M, et al. Vascular biomarkers and correlation with peripheral vasculopathy in systemic sclerosis. Autoimmunity Reviews. 2015;**14**:314-322

[34] Manfredi A, Sebastiani M, Cassone G, et al. Nailfold Capillaroscopi changes in dermatomyositis and polymyositis. Clinical Rheumatology. 2015;**34**:279-284

[35] Selva-O'Callaghan A, Fonollosa-Pla V, TRallero-Araguás E, et al. Nailfold capillary microscopy in adults with inflammatory miopathy. Seminars in Arthritis and Rheumatism. 2010;**39**(5):398-404

[36] de Holanda A, Bonfá E, Fuller R, et al. Capillaroscopy is a dynamic process in mixed connective tissue disease. Lupus. 2007;**16**:254-258

[37] Vayssairat M, Priollet P. Atlas pratique de capillaroscopie. 1ere éd ed. Paris: Les éditions de la revue de Médecine; 1983

[38] Ribeiro C, Holler A, Skaer T, et al. Periungual capillaroscopy in psoriasis. Anais Brasileiros de Dermatologia. 2012;**87**(4):550-553

[39] Bertolazzi C, Gallegos-Nava S, Villarreal-Treviño A, et al. The current role of capillaroscopy in vasculitides. Clinical Rheumatology. 2019;**38**(9):2299-2307. DOI: 10.1007/s10067-018-4399-1

[40] Jenette JC, Falk RJ, Bacon PA, et al. 2012 revised international Chapel Hill consensus conference nomenclature of vasculitides. Arthritis & Rheumatology. 2013;**65**:1-11

[41] Rossi D, Mansouri M, Baldovino S, et al. Nailfold videocapillaroscopy in mixed cryoglobulinaemia. Nephrology, Dialysis, Transplantation. 2004;**19**:2245-2249

[42] Movasat A, Shahram F, Carreira PE, et al. Nailfold capillaroscopy in Behçet's disease, analysis of 128 patients. Clinical Rheumatology. 2009;**28**:603-605

[43] Javinani A, Pournazari M, JAmshidi A, et al. Nailfold videocapillaroscopy changes in Takayasu arteritis and their association with disease activity and subclavian artery involvement. Microvascular Research. 2019;**122**:1-5

[44] Huang MY, Huang JJ, Huang TY. Deterioration of cutaneous microcirculation status of Kawasaki disease. Clinical Rheumatology. 2012;**31**:847-852

[45] Anders HJ, Haedecke C, Sigl T, et al. Avascular areas on nailfold capillary microscopy of patients with Wegener's granulomatosis. Clinical Rheumatology. 2000;**19**:86-88

[46] Greenberg LW. Nailfold capillary abnormalities in Henoch-Schönlein purpura. The Journal of Pediatrics. 1983;**103**:665-666

[47] Martino F, Agolini D, Tsalikova E. Nailfold capillaroscopy in Henoch-Schönlein purpura: A follow-up study of 31 cases. The Journal of Pediatrics. 2002;**141**:145

[48] Zampetti A, Rigante D, Bersani G, et al. Longitudinal study of microvascular involvement by nailfold capillaroscopy in children with

Henoch-Schönlein purpura. Clinical Rheumatology. 2009;**28**:1101-1105

[49] Vaiopoulos G, Pangratis N, Samarkos M. Nailfold capillary abnormalities in Behçet's disease. The Journal of Rheumatology. 1995;**22**:1108-1111

[50] Aytekin S, Yuksel EP, Aydin F, et al. Nailfold capillaroscopy in Behçet disease, performed using videodermoscopy. Clinical and Experimental Dermatology. 2014;**39**:443-447

[51] Alan S, Balkarli A, Tuna S. The nailfold capillaroscopy findings of Behçet's syndrome. Dermatologica Sinica. 2016;**34**:74-77

[52] Kuryliszyn-Moskal A, Bernacka K, Klimiuk PA. Circulation intercellular adhesion molecule 1 in rheumatoid arthritis – Relationship to systemic vasculitis and microvascular injury in nailfold capillary microscopy. Clinical Rheumatology. 1996;**15**:367-373

[53] Kuryliszyn-Moskal A. Cytokines and soluble CD4 and CD8 molecules in rheumatoid arthritis: Relationship to systematic vasculitis and microvascular capillaroscopic abnormalities. Clinical Rheumatology. 1998;**17**:489-495

[54] Witkowska AM, Kuryliszyn-Moskal A, Borawska MH, et al. A study on soluble intercellular adhesion molecule-1 and selenium in patients with rheumatoid arthritis complicated by vasculitis. Clinical Rheumatology. 2003;**22**:414-419

[55] Lambova S, Müller-Ladner U. Capillaroscopic features of microangiopathy in rheumatoid arthritis patients with peripheral vascular syndrome. Clinical Rheumatology. 2019;**38**(9):2339-2341. DOI: 10.1007/s10067-019-04561-x

[56] Rajaei A, Dehghan P, Amiri A. Nailfold capillaroscopy in 430 patients with rheumatoid arthritis. Caspian Journal of Internal Medicine. 2017;**8**(4):269-274

[57] Pyrpasopoulou A, Triantafyllou A, Anyfanti P, et al. Capillaroscopy as a screening test for clinical antiphospholipid syndrome. European Journal of Internal Medicine. 2011;**22**:e158-e159

[58] Sulli A, Pizzoni C, Cutolo M, et al. Nailfold capillaroscopy abnormalities in patients with antiphospholipid antibodies. The Journal of Rheumatology. 2000;**27**:1574-1576

[59] Maldonado G, Guerrero R, Paredes C, et al. Nailfold capillaroscopy in diabetes mellitus. Microvascular Research. 2017;**112**:41-46

[60] Hsu PC, Liao PY, Chang HH, et al. Nailfold capillary abnormalities are associated with type 2 diabetes progression and correlated with peripheral neuropathy. Medicine. 2016;**95**(52):e5714

[61] Gasser P, Bühler FR. Nailfold microcirculation in normotensive and essential hypertensive subjects, as assessed by video-microscopy. Journal of Hypertension. 1992;**10**:83-86

[62] He FJ, Marciniak M, Markandu ND, et al. Effect of modest salt reduction on skin capillary rarefaction in white, black and Asian individuals with mild hypertension. Hypertension. 2010;**56**:253-259

Chapter 12

Nailfold Capillaroscopy in Rheumatic Diseases

Abhishek Patil and Isha Sood

Abstract

Nailfold capillaroscopy (NFC) has developed into an indispensable tool for rheumatologists in the evaluation of rheumatic diseases. It offers various advantages in being rapid, noninvasive, and inexpensive. With NFC we are able to visualize the microcirculatory changes in the nail beds. These changes are key to the pathogenesis of connective tissue diseases such as systemic sclerosis. Hence NFC helps in early diagnosis of various connective tissue diseases. There is a lack of standardization in the techniques used and various capillary parameters studied, which could lead to variation in the reporting of the parameters studied. In this chapter we shall try to highlight the most common parameters studied in capillaroscopy and its utility in various connective tissue diseases.

Keywords: scleroderma, myositis, SLE, patterns

1. Introduction

Nailfold capillaroscopy (NFC) is a noninvasive bedside tool for the assessment of capillary microcirculation and its changes. The history of this technique dates back to over four centuries, with J C Kohlhaas's first description of capillary loops in nailfold using basic optical magnification. However, the utility of NFC in rheumatological diseases came to the forefront after the studies of Maricq and Le Roy in 1973 [1]. They systematically described the capillaroscopic patterns in patients of rheumatoid arthritis, systemic lupus erythematosus (SLE), and scleroderma (SSc) using a wide-field stereomicroscope of 12× magnification. From then on, NFC has evolved to be an indispensable tool for the rheumatologists in the evaluation of patients with connective tissue diseases. NFC is now included in the recent American College of Rheumatology/European League Against Rheumatism classification criteria for SSc [2]. In the current chapter, we shall highlight the physiology, preparation, instruments, and NFC parameters in various rheumatic diseases including SSc.

2. Principle

Microcirculation consists of arterioles, capillaries, and venules. The main function of the microcirculation is capillary exchange—delivery of oxygen and nutrients to tissues and removal of carbon dioxide and waste products. In most areas of fingers, capillaries are oriented perpendicular to the skin surface and are thus not amenable to visualization. In the nailfold areas, however, they become parallel to the skin surface and thus are observable in full length in the distal row.

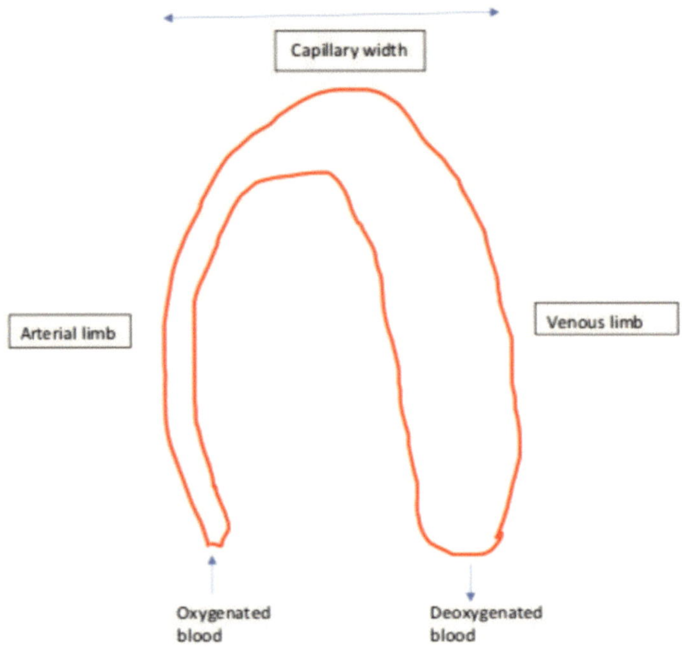

Figure 1.
Schematic representation of a capillary loop.

Usually there are one to three capillaries in each dermal papilla [3]. Hence most of the examination of the capillaries is concentrated on this region. Each capillary loop consists of an arterial and a venous limb with the latter being wider (**Figure 1**). With the help of various instruments, we are able to visualize the column of blood within these capillaries. The density and morphology of these loops help in the diagnosis and follow-up of patients with rheumatic diseases. We shall address these in the below mentioned sections.

3. Instruments used in capillaroscopy

The instruments utilized for the study of nailfold capillaries have evolved over time, from stereomicroscope to the high magnification nailfold video capillaroscopy (NVC).

3.1 Wide-field stereomicroscope

It is the original technique utilized by Maricq et al. [1] and employs about 20× magnification for the evaluation of capillaries. It provides the panoramic view of nailfold capillaries and makes possible for the assessment of qualitative and quantitative parameters (**Figure 2a**). However, the instrument is expensive and non-portable, and access is limited to special interest groups.

3.2 Ophthalmoscope and dermatoscope

These portable instruments provide images with lower magnification and quality. They are best utilized as bedside evaluation tools for the clinicians. Despite their lower cost, they lack the image storing and processing capabilities.

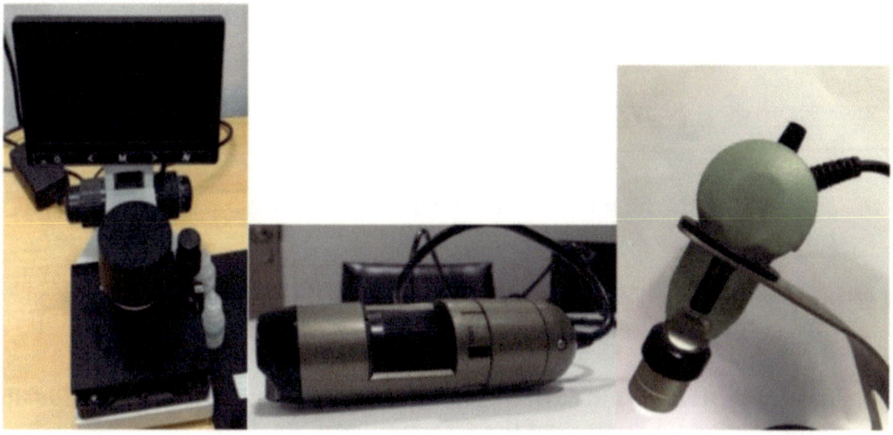

2a 2b 2c

Figure 2.
Various instruments used in nail fold capillaroscopy; (a) stereomicroscope, (b) microscope and (c) nail fold videocapillaroscope.

3.3 Microscope

It usually combines optical microscope and digital video camera connected to a computer (**Figure 2b**). It's a handheld, inexpensive tool for the evaluation of capillary parameters. The instrument can be used in varying magnifications from 200× to 600×. In view of its portability and low cost, its best suited to the busy outpatient settings.

3.4 Nailfold videocapillaroscope (NFVC)

This technique provides high magnification (200× to 600×) and, with the aid of specific software, allows a precise measurement of capillaroscopic parameters (capillary length, width, and density) (**Figure 2c**). The disadvantages include high cost and loss of panoramic view of the nailfold.

4. Procedure of nailfold capillaroscopy

The procedure followed for nailfold examination is similar for all of the above-mentioned instruments. The underlying principle is to minimize the variability in tissue perfusion due to the following testing conditions:

i. Patient is instructed to avoid smoking and caffeine for at least 6 h prior to the procedure.

ii. He/she should be explained about the procedure.

iii. He/she is allowed to get acclimatized to room temperature for about 15–20 min.

iv. The skin to be examined is cleaned using soap and water.

v. Then a drop of skin-friendly oils such as cedarwood oil, olive oil, peanut oil, etc. should be placed on the nailfold and observed under NFC. It should be used in optimum quantity, and too little or too much of the oil may result in suboptimal resolution.

vi. A series of 4–12 overlapping images are taken in each finger to complete the procedure.

vii. Conventionally the medial four fingers of both the hands excluding the thumb are evaluated.

viii. Fingers which have sustained a recent injury are excluded from the analysis [4].

ix. Usually the fourth and fifth fingers provide the maximum information due to the transparent nature of the skin in them.

x. The parameters evaluated include capillary density, morphology, and hemorrhages. The same are discussed in detail below.

5. Capillary parameters

5.1 Capillary shape

A normal capillary has a safety pin or inverted U appearance with an arterial and venous limbs. The venous arm is larger than the arterial arm. However, the capillary loops may exhibit morphological variations, such as crossed or meandering loops (with intertwining). Bushy or excessively tortuous capillaries are seen in scleroderma, mixed connective tissue disease (MCTD), and dermatomyositis.

5.2 Capillary density (CD)

It is the number of capillaries in 1 mm length of the distal row of each finger or toe. A capillary loop is considered to be in the distal row, if the angle between the apex of that capillary loop and the two adjacent capillary loops is more than 90° [5]. Capillary density of more than 9 is usually taken as normal, albeit age-related changes in the parameter [6]. A mean CD for each finger is calculated by analyzing four fields in the same.

5.3 Capillary width

Literature has used different parameters while reporting on capillary width. Some studies have reported on the width of the apical loop, arterial limb, and venous limb, while the others have reported on the whole width of capillary loop (**Figure 3**). Here we shall consider the whole width of capillary loop while describing the dimensions. The normal capillary loop diameter ranges from 25 to 50 μm in adults. A capillary width of more than 90 μm is generally taken as an enlarged loop [7]. Uniformly enlarged capillaries are typically seen in SSc, dermatomyositis, and mixed connective tissue disease. The term giant capillary is used to denote a capillary diameter of more than four times the normal size.

5.4 Intercapillary distance (ICD)

It is the distance between apical tips of two capillaries and varies between 96 and 166 μm [8]. Avascular areas are defined as the areas in which two or more

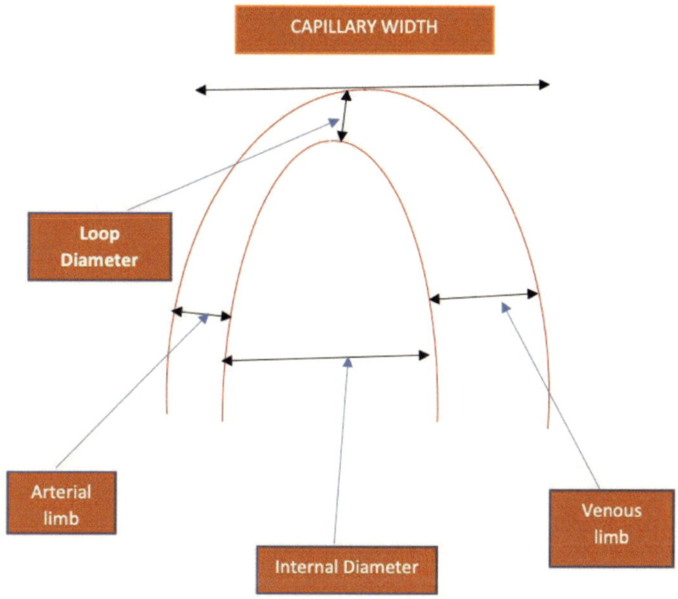

Figure 3.
Schematic representation of various capillary loop dimensions.

capillaries are missing as compared to the other areas with low capillary density [9]. Other authors term an ICD of more than 500 μm as avascular areas [8]. These are typically seen in connective tissue diseases and are not associated with nonimmune disorders.

5.5 Microhemorrhages

These are reddish brown punctate lesions found in the nail bed. These represent early vascular damage and are most prominently visualized in connective tissue diseases such as scleroderma.

6. Utility of capillary microscopy in rheumatology

Rheumatologists frequently encounter situations in which patients present with complaints such as arthralgia, Raynaud's, lung disease, etc. However, many of these patients lack the typical clinical features to classify as a definite systemic connective tissue disease such as scleroderma, MCTD, and others. In such scenarios, NFC can serve as a useful modality in differentiating patients with and without connective tissue disease. The common nail abnormalities encountered include dilated loops, microhemorrhages, and capillary dropouts. Here we shall study the NFC abnormalities encountered in patients with rheumatological diseases.

6.1 Raynaud's phenomenon

Raynaud's phenomenon is an exaggerated responsiveness of the vasculature to cold and other stimuli. It can be primary (in absence of an associated disorder) or secondary (associated with a systemic disease such as scleroderma). NFC can serve as a useful tool in distinguishing primary from secondary Raynaud's. Mannarino et al. [10] described three capillaroscopic patterns in patient with

Raynaud's—normal, borderline, and abnormal patterns. Authors proposed the borderline NFC abnormalities (composed of shorter and more tortuous capillaries) are due to the long-standing vasospastic reaction of the Raynaud's itself. Of interest though is the finding of abnormal capillaroscopic pattern in 8/44 patients with idiopathic Raynaud's phenomenon. These are the group of patients who might be at a higher risk for developing connective tissue disease on follow-up. On the other hand, the absence of NFC abnormalities in a patient of Raynaud's is also helpful in the exclusion of scleroderma [11].

7. Capillaroscopy in systemic sclerosis

Three capillaroscopic patterns have been described in SSc: early, active, and late [12].

SSc-early pattern: it is characterized by the presence of few giant capillaries, few capillary hemorrhages, relatively well-preserved capillary distribution, and no evident loss of capillaries (**Figure 4**).

SSc-active pattern: in active pattern frequent giant capillaries are seen with frequent capillary hemorrhages, moderate loss of capillaries, mild disorganization of the capillary architecture, and mild ramified capillaries (**Figure 5**).

SSc-late pattern: in late pattern irregular enlargement of the capillaries with few or absent giant capillaries and hemorrhages are seen. There is severe loss of capillaries with extensive avascular areas, disorganization of the normal capillary array, and ramified capillaries (**Figure 6**).

7.1 Nailfold capillaroscopy and correlation with organ involvement

There have been several studies which tried to look for correlation of capillaroscopic parameters with organ involvement in systemic sclerosis. Though the results from these studies have been variable, marked skin involvement and the presence

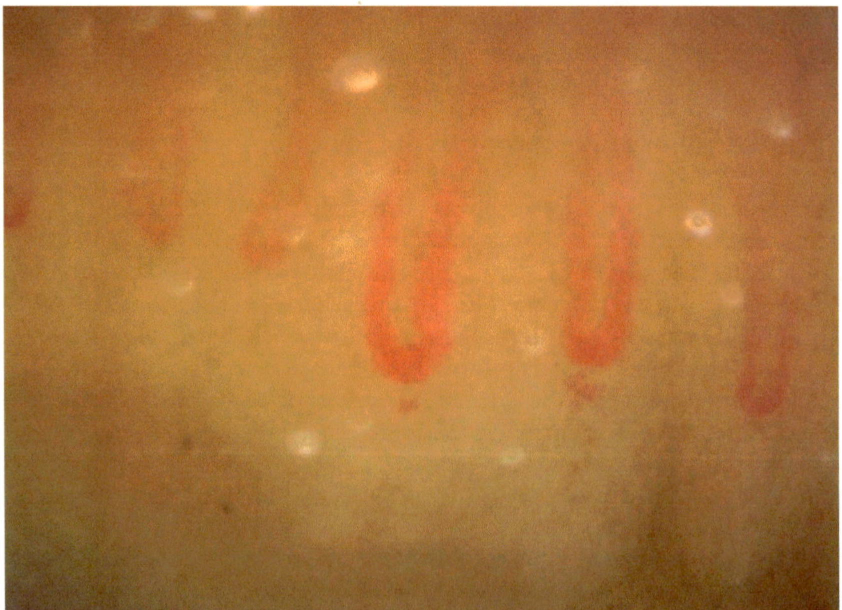

Figure 4.
Early SSc pattern showing few giant capillaries.

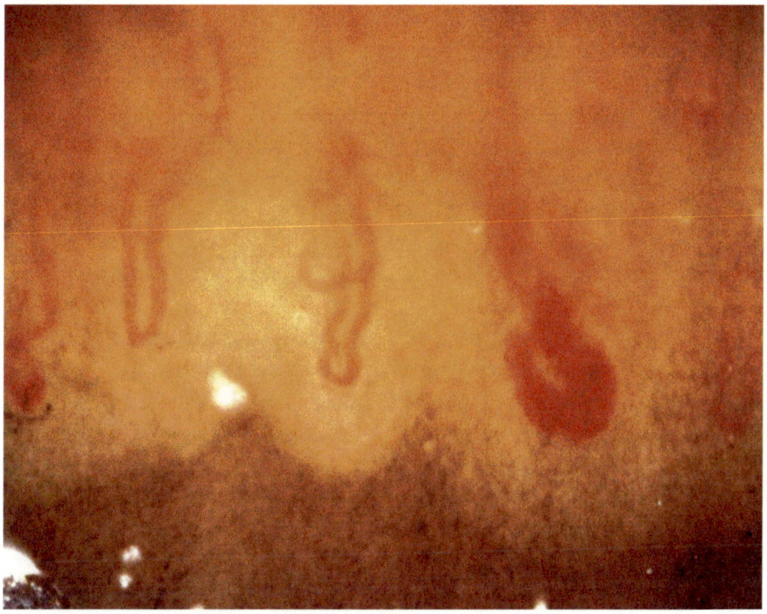

Figure 5.
Active SSc pattern showing hemorrhages and giant capillaries.

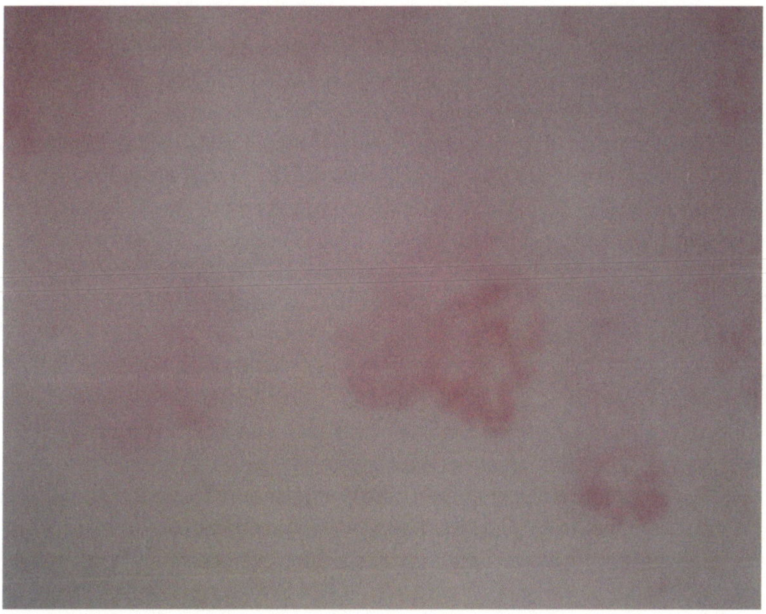

Figure 6.
Late SSc pattern showing severe architectural distortion and capillary dropouts.

of pulmonary arterial hypertension were found to correlate with capillary loss in majority of these studies.

Sato et al. [13] and Bhakuni et al. [14] found that capillary loss was associated with higher skin scores as assessed by modified Rodnan skin scores. Also the significant loss of capillaries in diffuse cutaneous disease as compared to limited disease was observed by Bhakuni et al. [14] and Ostojic et al. [15] in their study. These studies showed, with the increasing skin thickness, the capillary density decreases and thus patients with diffuse cutaneous disease which has higher skin scores have more marked capillary loss.

The association between the presence of pulmonary hypertension (PAH) and various capillary parameters was also studied using either echocardiography or right heart catheterization to define PAH. The first study of NFC to use right heart catheterization (RHC) evaluated 44 SSc patients and found that SSc patients with more capillary abnormalities (defined by increased apical limb width, capillary width, area, and capillary length) correlated with higher pulmonary vascular resistance, but capillary density was not evaluated in this study [16]. Hofstee et al. [17] in a recent study using RHC (mean pulmonary artery pressure (PAP) of >25 mmHg at rest or >30 mmHg during exercise as cutoff for PAH) found capillary loss to be significantly associated with PAH. Riccieri et al. [18] used echocardiography for screening (PASP>35) and RHC for confirmation of PAH. A total of 24 patients of SSc were studied of which 12 had PAH. Significantly more capillary alterations and more avascular areas were found in patients with PAH. Among the studies using echocardiographic screening for PAH, Sato et al. [13] defined PAH as PASP >35 mmHg and could not find any significant association between the presence of PAH and various capillary parameters. Castellvi et al. however found the neoangiogenesis to be significantly associated with PAH (PASP > 40 on echocardiography) [19]. One more study, did not find significant difference among NFC patterns in patients with elevated PASP [20].

The correlation between the presence of interstitial lung disease and capillaroscopy parameters is not as consistent as that for PAH. Castellvi et al. [19] found that patients with loss of capillaries on NFC had worse DLCO and FVC; however none of the three SSc patterns (early, active, and late) showed any association with FVC/DLCO ratio. Another study using FVC < 75% of predicted value and/or HRCT and/or chest radiographic changes to define ILD could not find any difference in capillary parameters in patients with and without ILD [20]. However 1 study which screened 91 patients with SSc did found that patients with ground glass opacities on HRCT had significantly higher mean avascular scores [21]. Another recent study found that degree of neoangiogenesis was higher in SSc patients with honeycombing and DLCO < 50% and a number of avascular areas inversely correlated to DLCO/AV (alveolar volume) [22].

Among serologies some studies could find an association between Scl-70 and capillary density while no association was found with anti-centromere. Anti-Scl-70 positivity is seen in around 40% of dCSSc patients and carries an increased risk of mortality owing to its association with lung fibrosis and rapidly developing skin thickness [23]. The association of Scl-70 with active and late pattern was reported for the first time by Cutolo et al. [24]. Anti-Scl-70 antibodies were found to be significantly more prevalent in those with the "late" scleroderma pattern of capillaroscopy than "active" and "early" pattern in the EUSTAR cohort [25]. Another study found that vascular deletion score was significantly higher in patients with anti-Scl-70 positivity; however no correlation with capillary density was found [21]. A recent study could not find significant difference in NFC parameters between Scl-70 positive and negative patients, though avascular areas were numerically higher [22].

Association of anti-centromere antibodies with any of the parameter on capillaroscopy have been inconsistent. Herrick et al. found that reduced capillary density was associated with positive anti-centromere antibody [26]. Another study found that anti-centromere was significantly more prevalent in early and active pattern on capillaroscopy than late scleroderma pattern [27]. Most of the other recent studies failed to find significant relationship between ACA positivity and

NFC parameters [13, 22]. The presence of ACA has been associated with a lower frequency and severity of radiographic interstitial pulmonary fibrosis [28].

8. Nailfold capillaroscopy in other rheumatic diseases

The role of NFC in other rheumatic diseases is much well established compared to SSc. Some of the salient features of NFC in various rheumatic diseases are discussed below.

8.1 Systemic lupus erythematosus (SLE)

Nearly half of the patients with SLE have a normal capillaroscopic pattern. Others have a tortuous, meandering capillaries, bizarre loops, and a prominent subcapillary venous plexus [29]. In a systematic review by Cutolo et al. [30], 40 studies describing the capillaroscopic patterns in SLE were studied. They found the meandering capillaries and hemorrhages are to occur more frequently in patients with SLE and significantly less hairpin shaped loops compared to healthy individuals. Of note, they found the dilated capillaries to be associated with Raynaud's and gangrene in lupus patients. In seven of these studies, NFC scores also correlated with the disease activity.

8.2 Dermatomyositis (DM) and polymyositis

The scleroderma pattern is observed in 20–60% of patients with more frequent and pronounced findings in dermatomyositis [31]. Juvenile dermatomyositis (JDM) can reveal phasic changes on NFC. Early stages of microhemorrhages and giant capillaries are followed later by capillary loss and neoangiogenesis [32]. In dermatomyositis, these phasic changes are less obvious. Interestingly, shorter duration of disease is associated with more sever changes—giant capillaries and reduced capillary density. Longer duration of disease is typified by the presence of extensively ramified capillaries [33]. There also appears to be a strong correlation between the NFC involvement and cutaneous activity [34]. In JDM patients, lack of resolution of NFC changes is associated with the more severe and chronic forms of the disease [32]. An association has also been described between capillary abnormalitics and ILD, Raynaud's, and malignancy [35]. However, currently the data correlating with disease activity is sparse to formulate any definitive recommendations.

8.3 Mixed connective tissue disease (MCTD)

Scleroderma pattern is observed in 65% patients of MCTD. The presence of avascular areas has a strong correlation with the presence of interstitial lung disease [36]. In a proportion of the patients, the NFC changes can revert with normalization of CD and/or improvement in dilated loops and hemorrhages [36].

8.4 Undifferentiated connective tissue disease (UCTD)

NFC can help in the establishing the presence of a connective tissue disease in a patient with equivocal findings. Scleroderma pattern is observed in 9/65 (13.8%) of patients with UCTD in one study [37]. Approximately a quarter of the patients with UCTD may transform into SSc on follow-up [38]. Hence capillaroscopic examination could be of value in identifying the patients of UCTD who could progress to SSc on follow-up.

8.5 Sjogren's syndrome (SjS)

Patients of SjS who have Raynaud's phenomenon have crossed capillaries and confluent hemorrhages on NFC [39]. Also the patients who had anti-centromere antibodies had scleroderma-type findings in this particular study.

9. Conclusions

There are two principal methods utilized in nailfold capillaroscopy. The one with the stereomicroscope provides the panoramic view at lesser magnification. Videocapillaroscope provides a larger magnification but at the cost of limited field of visualization. They are equivalent for the identification of classical abnormalities. The inclusion of capillaroscopic abnormalities in the ACR/EULAR classification for SSc drew more attention to this technique. The education and training of rheumatologists in capillaroscopic examination is the need of the hour.

Author details

Abhishek Patil[1*] and Isha Sood[2]

1 DNB (Rheumatology), Manipal Hospitals, Bangalore, India

2 DNB (Rheumatology), Nanavati Superspeciality Hospital, Mumbai, India

*Address all correspondence to: abhishekpatilkims@gmail.com

References

[1] Maricq HR, LeRoy EC. Patterns of finger capillary abnormalities in connective tissue disease by "wide-field" microscopy. Arthritis and Rheumatism. 1973;**16**:619-628

[2] van den Hoogen F, Khanna D, Fransen J, et al. Classification criteria for systemic sclerosis: An ACR-EULAR collaborative initiative. Arthritis and Rheumatism. 2013;**65**(11):2737-2747

[3] Schiavon F, Maffei P, Martini C, et al. Morphologic study of microcirculation in acromegaly by capillaroscopy. Journal of Clinical Endocrinology & Metabolism. 1999;**84**:3155-3159

[4] Hoerth C, Kundi M, Katzenschlager R, Hirschl M. Qualitative and quantitative assessment of nailfold capillaries by capillaroscopy in healthy volunteers. VASA. 2012;**41**:19-26

[5] Hofstee HMA, Serne EH, Roberts C, et al. A multicentre study on the reliability of qualitative and quantitative nail-fold videocapillaroscopy assessment. Rheumatology. 2012;**51**:749-775

[6] Cutolo M, Sulli A, Smith V. How to perform and interpret capillaroscopy. Best Practice & Research. Clinical Rheumatology. 2013;**27**(2):237-248

[7] Dolezalova P, Young SP, Bacon PA, South-Wood TR. Nailfold capillary microscopy in healthy children and in childhood rheumatic diseases: A prospective single blind observational study. Annals of the Rheumatic Diseases. 2003;**62**(5):444-449

[8] Etehad Tavakol M, Fatemi A, Karbalaie A, Emrani Z, Erlandsson BE. Nailfold capillaroscopy in rheumatic diseases: Which parameters should be evaluated? BioMed Research International. 2015;**2015**:974530

[9] Schiavon F, Maffei P, Martini C, et al. Morphologic study of microcirculation in acromegaly by capillaroscopy. The Journal of Clinical Endocrinology and Metabolism. 1999;**84**(9):3151-3155

[10] Mannarino E, Pasqualini L, Fedeli F, Scricciolo V, Innocente S. Nailfold capillaroscopy in the screening and diagnosis of Raynaud's phenomenon. Angiology. 1994;**45**(1):37-42

[11] Bissell LA, Abignano G, Emery P, Del Galdo F, Buch MH. Absence of scleroderma pattern at nail fold capillaroscopy valuable in the exclusion of scleroderma in unselected patients with Raynaud's phenomenon. BMC Musculoskeletal Disorders. 2016;**17**(1):342

[12] Cutolo M, Sulli A, Pizzorni C, Accardo S. Nailfold videocapillaroscopy assessment of microvascular damage in systemic sclerosis. The Journal of Rheumatology. 2000;**27**:155-160

[13] Sato L, Kayser C, Andrade L. Nailfold capillaroscopy abnormalities correlate with cutaneous and visceral involvement in systemic sclerosis patients. Acta Reumatológica Portuguesa. 2009;**34**:219-227

[14] Bhakuni DS, Vasdev V, Garg MK, Narayanan K, Jain R. Nailfold capillaroscopy by digital microscope in an Indian population with systemic sclerosis. International Journal of Rheumatic Diseases. 2012;**15**:95-101

[15] Ostojić P, Damjanov N. Different clinical features in patients with limited and diffuse cutaneous systemic sclerosis. Clinical Rheumatology. 2006;**25**:453-457

[16] Ohtsuka T, Hasegawa A, Nakano A, Yamakage A, Yamaguchi M, Miyachi Y. Nailfold capillary

abnormality and pulmonary hypertension in systemic sclerosis. International Journal of Dermatology. 1997;**36**:116-122

[17] Hofstee HMA, Noordegraaf AV, Voskuyl AE, Dijkmans BAC, Postmus PE, Smulders YM, et al. Nailfold capillary density is associated with the presence and severity of pulmonary arterial hypertension in systemic sclerosis. Annals of the Rheumatic Diseases. 2009;**68**:191-195

[18] Riccieri V, Vasile M, Iannace N, Stefanantoni K, Sciarra I, Vizza CD, et al. Concise report systemic sclerosis patients with and without pulmonary arterial hypertension: A nailfold capillaroscopy study. Rheumatology. 2013;**52**:1525-1528

[19] Castellví I, Simeón-aznar CP, Sarmiento M, Fortuna A, Geli C, Diaz-torné C, et al. Association between nailfold capillaroscopy findings and pulmonary function tests in patients with systemic sclerosis association between nailfold capillaroscopy findings and pulmonary function tests in patients with systemic sclerosis. The Journal of Rheumatology. 2015;**42**:222-227

[20] Caramaschi P, Canestrini S, Martinelli N, Volpe A, Pieropan S, Ferrari M, et al. Scleroderma patients nailfold videocapillaroscopic patterns are associated with disease subset and disease severity. Rheumatology. 2007;**46**:1566-1569

[21] Bredemeier M, Xavier RM, Capobianco KG, Restelli VG, Rohde LE, AFF P, et al. Nailfold capillary microscopy can suggest pulmonary disease activity in systemic sclerosis. The Journal of Rheumatology. 2004;**31**:286-294

[22] De Santis M, Ceribelli A, Cavaciocchi F, Crotti C, Massarotti M, Belloli L, et al. Nailfold videocapillaroscopy and serum VEGF levels in scleroderma are associated with internal organ involvement. Autoimmunity Highlights. 2016;7:1-9

[23] Spencer-Green G, Alter D, Welch H. Test performance in systemic sclerosis: Anti-centromere and anti-Scl-70 antibodies. The American Journal of Medicine. 1997;**103**:242-248

[24] Cutolo M, Pizzorni C, Tuccio M, Burroni A, Craviotto C, Basso M, et al. Nailfold videocapillaroscopic patterns and serum autoantibodies in systemic sclerosis. Rheumatology. 2004;**43**:719-726

[25] Ingegnoli F, Ardoino I, Boracchi P, Cutolo M, Airò P, Ananieva LP, et al. Nailfold capillaroscopy in systemic sclerosis: Data from the EULAR scleroderma trials and research (EUSTAR) database. Microvascular Research. 2013;**89**:122-128

[26] Herrick AL, Moore TL, Murray AK, Whidby N, Manning JB, Bhushan M, et al. Original article nail-fold capillary abnormalities are associated with anti-centromere antibody and severity of digital ischaemia. Rheumatology. 2010;**49**:1776-1782

[27] Shenavandeh S, Haghighi MY, Nazarinia MA. Nailfold digital capillaroscopic findings in patients with diffuse and limited cutaneous systemic sclerosis. Rheumatology. 2017;**1**:23-31

[28] Ho KT, Reveille JD. The clinical relevance of autoantibodies in scleroderma. Arthritis Research & Therapy. 2003;**5**:80-93

[29] Kabasakal Y, Elvins DM, Ring EF, McHugh NJ. Quantitative nailfold capillaroscopy findings in a population with connective tissue disease and in normal healthy controls. Annals of the Rheumatic Diseases. 1996;**55**:507-512

[30] Cutolo M, Melsens K, Wijnant S, et al. Nailfold capillaroscopy in

systemic lupus erythematosus: A systematic review and critical appraisal. Autoimmunity Reviews. 2018;**17**(4):344-352

[31] Ganczarczyk ML, Lee Armstrong SK. Nailfold capillary microscopy in polymyositis and dermatomyositis. Arthritis and Rheumatism. 1988;**31**:116-119

[32] Bertolazzi C, Cutolo M, Smith V, Gutierrez M. State of the art on nailfold capillaroscopy in dermatomyositis and polymyositis. Seminars in Arthritis and Rheumatism. 2017;**47**(3):432-444

[33] Manfredi A, Sebastiani M, Cassone G, Pipitone N, Giuggioli D, Colaci M, et al. Nailfold capillaroscopic changes in dermatomyositis and polymyositis. Clinical Rheumatology. 2015;**34**:279-284

[34] Christen-Zaech S, Seshadri R, Sundberg J, Paller AS, Pachman LM. Persistent association of nailfoldcapillaroscopy changes and skin involvement over thirty-six months with duration of untreated diseases in patients with juvenile dermatomyositis. Arthritis and Rheumatism. 2008;**58**:571-576

[35] Mugii N, Hasegawa M, Matsushita T, Hamaguchi Y, Horie S, Yahata T, et al. Association between nailfold capillary findings and disease activity in derma-tomyositis. Rheumatology. 2011;**50**:1091-1098

[36] de Holanda Malfado Diógenes A, Bonfá E, Fuller R, Correia Caleiro MT. Capillaroscopy is a dynamic process in mixed connective tissue disease. Lupus. 2007;**16**:254-258

[37] Nagy Z, Czirjác L. Nailfold digital capillaroscopy in 447 patients with connective tissue disease and Raynaud's disease. Journal of the European Academy of Dermatology and Venereology. 2004;**18**:62-68

[38] Ohtsuka T. Quantitative analysis of nailfold capillary abnormalities in patients with connective tissue diseases. International Journal of Dermatology. 1999;**38**(10):757-764

[39] Tektonidou M, Kaskani E, Skopouli FN, Moutsopoulos HM. Microvascular abnormalities in Sjögren's syndrome: Nailfold capillaroscopy. Rheumatology (Oxford, England). 1999;**38**(9):826-830

Chapter 13

Basic Principles in Microvascular Anastomosis and Free Tissue Transfer

Ignacio Vila, Iván Couto-González and Beatriz Brea-García

Abstract

Free tissue transfer pursues the best functional and aesthetic results in reconstructive surgery. As these techniques completely maximise the donor tissues' disposability, these treatments have become a first-line option in many situations. When the donor site is taken form the same patient, these surgeries are often referred to as autotransplants. Free tissue transfer sustains in microvascular anastomosis, which are defined by a vessel lumen diameter inferior to 3 mm. Particular attention to some details is important in these techniques, as, for example, to preclude any damage to the vessel walls or any leakage in the microvascular anastomosis. But the success of these techniques does not only depend on an adequate vascular suture, but also on a constellation of details that must be taken into account. These go from the availability of a trained team, to the ergonomics of the surgeon, through the scrupulous cleanliness of the surgical field.

Keywords: free tissue transplantation, microvascular anastomosis, microsurgery, reconstruction

1. Introduction to microsugery

Microvascular transfer is a reconstructive technique based on raising tissues from healthy areas of the body, where an excess or dispensability exists, in advance to transplant them to other regions where they are lacking, mainly after trauma, oncological surgery or chronic infection. A microsurgical transfer from a strict point of view implies a double vascular anastomosis less than 3 mm between vessels in the transferred tissue to the ones in the recipient area [1]. Super-microsurgery would refer to those situations in which anastomoses have a diameter between 0.3 and 0.8 mm [2]. Rigorously speaking, the recipient vessels are those receiving the blood flow and the donors those from which it emanates. From a historical point of view, the compound of the transferred tissues is named free flaps.

Since its inception, reconstructive techniques have aimed to restore the integrity, form and function of the body [3]. Although plastic surgery is the discipline of medicine that brings together all these techniques, it lacks an anatomical limitation; therefore, its knowledge is widespread according to the diverse body regions through maxillofacial surgery, ophthalmology, hand surgery, etc. For centuries, it was intended to limit the potential damage inflicted to patients by narrowing down the reconstructive options. In this regard, a reconstructive ladder was defined, where the primary closure of the wounds, the cure by secondary intention or the skin grafts were in the lower steps of this ladder and the flaps in the higher [4, 5].

With the improvement in optical tools, it became easier to perform the vascular anastomoses that allowed free flap transfer and to set up skilled teams. As the tissue transfers became more dynamic and the microsurgery success rates rose, the benefits became more and more evident [5]. It was proven that the transfer of healthy tissues to the hand or head and neck allowed surgeons to achieve faster and better recoveries in areas of high functional demand, also with much more aesthetically acceptable results and lower morbidity. The same happened to breast surgery, where reconstructions with a natural shape and adequate volume could be achieved; the scars were hidden in the distance, and there was no need to use prosthesis. In lower limb osteomyelitis, free muscle flaps became the alternative to amputation. In addition, the advent of perforator flaps, mainly due to the contributions of Song and Koshima [5], thanks to whom it was not necessary to take the underlying muscle to transfer a fasciocutaneous flap, made it possible to further minimise the morbidity of these microsurgical interventions. Finally, a revision of the reconstructive ladder was proposed, the simplicity of the reconstruction would prevail, but pursuing the best aesthetic and functional results. So, a switch to a reconstructive elevator was made. In this way, microsurgical reconstructions became the first-line option for many patients and the technique was extended to a multitude of centres [5, 6].

2. Basic principles in microsurgery

2.1 Ergonomy

Multiple aspects regarding the environment in the operating room and the position are particularly important in microsurgery. It is imperative to have enough field to allow an easy movement. This aspect, which is less substantial in *macro*-surgery, becomes absolutely fundamental in microsurgery. Mention it at the beginning, does nothing but tries to emphasise its relevance.

A two-team approach is usually chosen in reconstructive microsurgery, one will raise the flap and the other will set the recipient site where this is going to be transplanted [7]. Therefore, all the time spent planning disposition is properly invested. This is true both for placing the patient in the proper position, and for the surgeon to adopt a comfortable and durable posture. Since the surgery will be prolonged, we must meticulously paddle all bony prominences of the patient and the areas at risk of neurovascular compression. It must be encouraged to take all the necessary anaesthesia monitoring measures at the beginning, just to avoid emergencies or interruptions during delicate stages of surgery. It is also sensible to foresee how the microscope will be arranged in the room.

The comfort of the surgeon is a must when it comes the time to perform the microvascular anastomosis, primarily regarding the back, scapular and muscular groups. The sutures used usually size about 75–100 μm and the vessel lumen just a few millimetres; therefore, any tremor will greatly hinder the precision and success of the anastomoses. We cannot afford mistakes at any point of the microvascular anastomosis. The surgeon must be perpendicular disposition to the vessels and seated in a self-regulating chair that allows a self-sufficient height adjust. He or she should also be with the feet on a flat surface, the arms supported on a cloth and the hands on some comfortable place of the field to work only with the intrinsic muscles of the hand [1].

Patience is the cornerstone of microsurgery, calm dissection with no external worries or hurry [8]. For this to be the case, it is essential to be in an easy environment without any tensions among the team members. Fatigue will appear mainly at the most complex moments, well in the middle of long interventions. So, if we do not foresee a comfortable environment with all these details, which may seem insignificant at first, as soon as the least complication appears, the reconstruction will be at high risk. In the case of microsurgical reconstructions, comfort is not a luxury but a must.

2.2 General conditions

After having invested enough time planning the operating room configuration, it is time to choose the vessels in the recipient area, since those of the flap are already determined and are assumed to be healthy because of their undamaged origin. It is essential to emphasise that the dissection must be very scrupulous, some groups advocate applying tension to the tissues around the vessel, without any direct pulling or forceps grasping on it, as not to generate any intimal traumas that may cause a thrombotic source [8, 9]. Any injury to the intima of the vessel, unnoticed or not, will expose the subendothelial collagen of the lumen, leading to a thrombotic focus. There are situations where it is impossible not to manipulate the vessel, as it happens in cervical dissections for oncological reasons; in these cases, a high incidence of thrombosis in the recipient vein has been demonstrated [6].

We must choose healthy vessels, without excessive fibrotic or irradiated tissue around them, this will allow us to perform a clean dissection, achieving a blood-less field. If blood accumulates in the field, we should spare no expense in abundantly rinse the area and review haemostasis. Blood has a red light refraction that deteriorates the sight with usual optical tools and releases procoagulant factors inducing vascular thrombosis [10, 11]. In limbs with previous surgeries or trauma, in case of doubt, we must carry out explorations such as angiography or Doppler, to check the availability of adequate vessels [12–15]. We should recruit as much vessel length as necessary to prevent any tension in the anastomosis, since the use of vein grafts, although may be needed, should be avoided due to its higher incidence of complications.

Before sectioning the donor artery to which we are going to transfer our flap, we must ensure that it has a good flow, we should ideally evince pulse [8]. Once sectioned, it will only be valid if we observe the exit of an abundant spurt of pulsatile blood. On the other hand, the vein that receives the blood from the flap in the recipient area must have at least the diameter that the vein of the flap has; otherwise, a bottleneck will form and prevent a good return and a venous congestion may develop in the flap.

Once the vessels in the receiving area are all set, we proceed to review the haemostasis and the perfusion of our previously dissected flap, then we release and transfer it [8, 10]. We should section the artery first and then the vein, as to avoid any congestion. Then we have to adapt the flap in the recipient area, since after anastomosis the flap will become edematised and its fixation in some deep spaces will be complex. This fixation is a mandatory prior step in all free flaps but in those in which the anastomosis lies in a deeper plane. In the head and neck reconstruction, small and intricate spaces make it advisable to do the fixation at first; but in breast reconstruction, we can only secure it with a gauze before microvascular anastomosis [6].

When performing the anastomosis, we prefer to adjust each vessel end in a simple microvascular clamp, tension-less approximate both ends and perform the microvascular anastomosis sparing as much proximal dissection as possible between the vessels of the flap pedicle and between the ones of the recipient area. On the other hand, we can place the anastomosis vessels end in a double microvascular clamp and approximate them [16]. The anastomosis should be placed on a rubber contrast and this over a wet gauze to avoid pooling and elevate the anastomosis from the surrounding field, full of thrombogenic debris [10] (**Figures 1** and **2**).

It is characteristic of the lumen of vessels to show a diameter smaller than that seen before severing them. This phenomenon is known as vasospasm. To mitigate it, we must perform a mechanical dilatation of the vessel lumen with specific dilator forceps and with agents such as lidocaine 1–2% or papaverine 3%, the latter being our preference [8–11]. Another dilatation technique is to abundantly rinse lumen with heparinised serum (200–300 IU/ml) [11–17]. It is key to remove the adventitia next to the anastomosis; we usually remove 2–3 mm with cutting technique, by pulling the adventitia over the lumen of the vessel and making a section parallel to the light. Aggressive adventitectomies leave the proximity of the anastamosis lacking *vasa vasorum*; this can cause ischemia in the vessel wall and, secondary to this, a failure or a pseudoaneurysm. On the other hand, the adventitia is highly thrombogenic, its entry with a knot into the lumen can be disastrous [18]. No technique completely removes the adventitia, but the sharp dissection seems more respectful with the intima [19]. Before carrying out the anastomosis, we must ensure that there are no intimal lesions in the lumen of the vessel, venous valves or branches, that may cause turbulence or resistance to flow in the vicinity of anastomosis [8].

There is debate about which anastomoses to perform first, whether arterial or venous. If there is no limitation for the position of a vessel deeper than another, as happens in breast reconstruction where the internal mammary vein usually has a more medial position, we can choose any one of them [6, 8]. Many groups choose to start with the arterial anastomosis to minimise isquemia time, taking into account that they do not usually experience added venous congestion. We usually start with the venous anastomosis to avoid any congestion within the flap that can cause a thrombus in its internal circuit. At the time of removing the clamps, once the anastomosis is completed, it is clearly preferable to remove the venous one first.

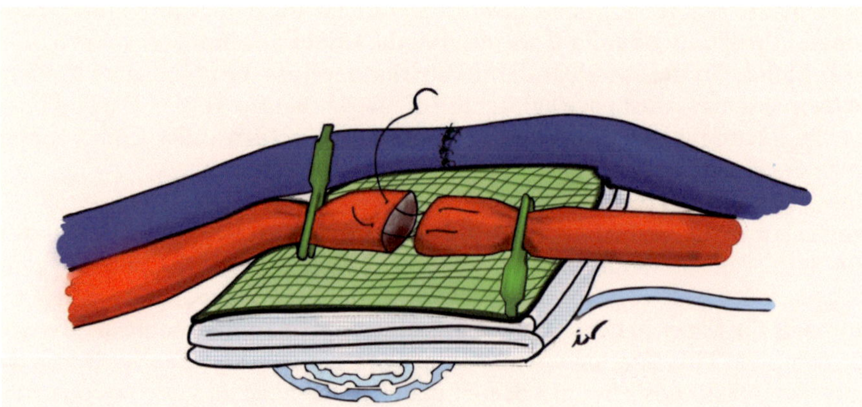

Figure 1.
General overview of a microvascular anastomosis. Artery microvascular anastomosis is performed in a blood-less field after vein anastomosis (in a second plane), in a higher position over a rubber medium contrast and wet gauzes. A protected 5F Redon is usually placed under the anastomosis. Dissection of the vessels in each side of the anastomosis is limited, only simple vascular clamps are employed; this eases the one-side-up technique (see below). In this figure, the first two stitches of the triangulation technique are depicted.

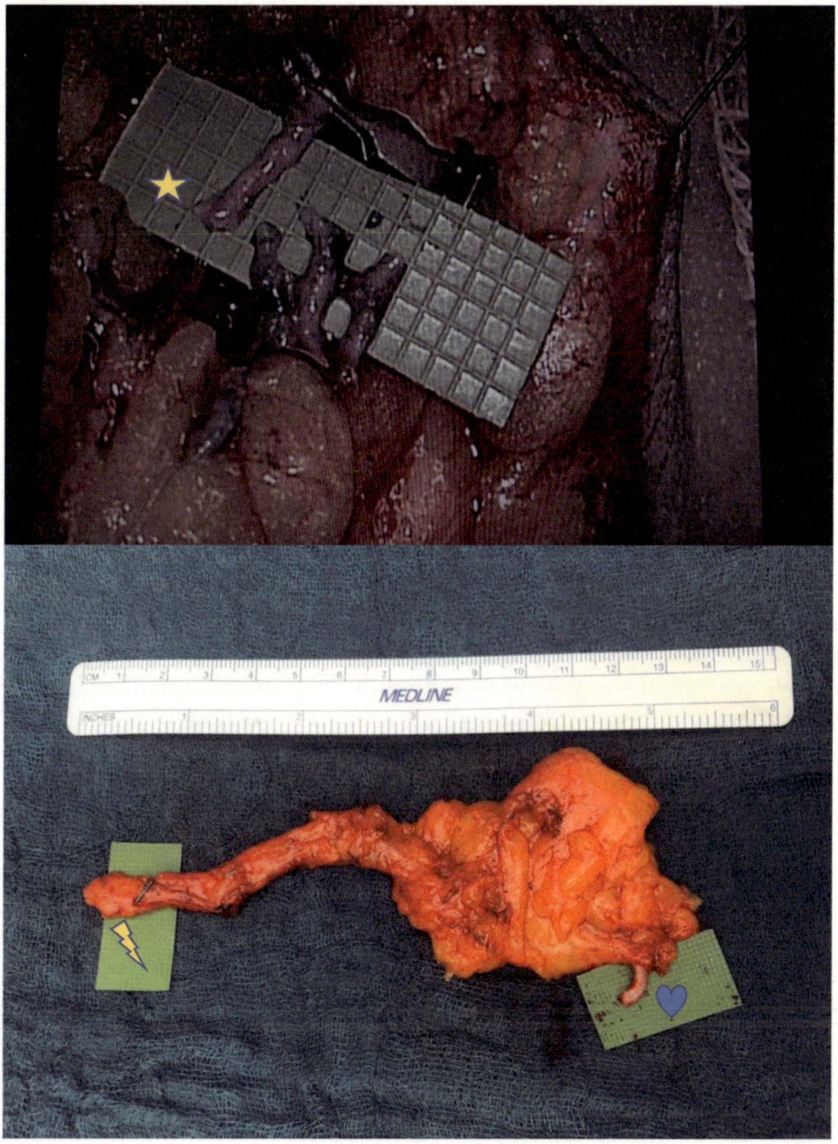

Figure 2.
Examples of the limited pedicle dissection before anastomosis. In the upper picture, we can see a free epigastric flap for an axillary free lymph node transfer. Marked with a star, we see the donor posterior circumflex humeral artery; below it, the inferior epigastric artery of the free flap is shown with both veins at each side. In the lower picture, we see the free flap before the transfer. Marked with a thunder, we see the inferior epigastric pedicle severed before its entrance in the abdominal rectus muscle; the inferior epigastirc vein is marked with a heart, it has been cut near its mouth in the circumflex iliac vein and laid prepared for the lymphovenous anastomosis.

2.3 Team

The use of microsurgical techniques is not limited to reference centres with a high availability of resources, although their routine use is almost exclusive of these. This is due to the disposability of a microsurgical team with several surgeons trained in microvascular anastomosis and free flaps management.

In this kind of surgery, each mistake has its consequences. So, if we perform these interventions with a very scarce and inexperienced team, these day-long surgeries can be translated into fatigue, nerve-wrecking and inaccuracies. This ultimately will generate failures in the microvascular anastomosis and problems in the perfusion of the flap. Therefore, having a team that allows pauses and relays, without stopping the procedure, is a fundamental element. Likewise, this second fresh team will overcome emergencies (more frequent in the first 48–72 h) or can replace a tired first team. It seems sensible to have at least four microsurgeons, two assistants and two experienced scrub nurses [20].

A microsurgical team must function as a unit that critically analyses its results, seeking rates of failure lower than 5% in free flaps. Errors and the morbidity of the interventions must be analysed, minimising both. This constant improvement is hard to achieve if several microsurgeons are not available.

3. Microsurgical tools and instruments

3.1 Tools

Microsurgery results from adapting the visual inaccuracy of our naked eye, to the fine movements of our hands. Here is where magnification arises. This can be done by two optical tools: the microscope and the magnifying loupe. In both, good lighting is essential [21, 22].

Surgical microscopes occupy a large space in the operating room but allow a magnification of up to 40× with greater illumination. In addition, they have pedals to control zoom and focus with the feet. Smaller magnification of 6–12× is usually used for the preparation of the vessels, and then it is increased up to 20× before the microanastomosis. In addition, the microscope gives us a wide range of field and provides the same vision to the surgeon and the assistant, enhancing the collaboration between them [21].

On the other hand, loupes are very cost-effective and easily transportable visual systems. The most common magnification employed in microsurgery is between 2.5× and 4.5×. In skilled hands, microscope has not proved to be superior *versus* loupes in achieving high success rates in free tissue transfers [23].

There are two types of magnifying loupes, on the one hand the compound or Galilean loupes and on the other hand the prismatic. The former consist of two lenses in line, and offer less weight and cost, although their magnification (2.5×) and depth of field are lower. The latter use a prism inside to reach a longer path of light through the lenses, which allows greater magnification and field depth, although they can be darker, heavier, more expensive and fragile [22].

3.2 Basic microsurgery instruments kit

The microsurgical instruments have evolved from ophthalmology or jewellery material to extremely specific and precise tools [7]. The basic kit is not made up of too much surgical material. This material should ideally be antireflective and cylindrical to allow its sliding from the index to the middle fingers and facilitate the passage of the needle through the tissues using only the intrinsic musculature. The size of the material should be about 16 cm to facilitate its support in the first hand commissure. In the case of working in very small fields, as in the case of hand surgery, smaller material, about 8 cm, with flat surface may be useful. Nowadays the self-locking material has lost interest, the mere requisite is just to offer little resistance when grasping to preclude any fatigue of the thenar eminence with prolonged use [24].

The basic kit consists of two scissors, a needle holder and a jeweller forceps. One of the scissors should be curved and round tipped, to be useful to dissect.

Other pair should be straight and pointed to perform the adventitectomy and to cut sutures. These pointed scissors should not be used for tissue dissection, because of the possible vessel trauma that they would generate. The jeweller forceps must have a precise closure, with enough contact between surfaces at the tip, just to handle fine sutures of 75 or 100 μm.

Other instruments that can also be useful are an aspiration system, an irrigation system and a bipolar forceps identical to jeweller forceps but protected. Our preference is to prepare a fixed suction system in the corner of the field, and to avoid introducing traditional aspirators directly over the vessels. Usually we fix a 5F Redon drain in a corner of the field or under the rubber contrast and we keep it connected to soft aspiration, in such way that it rests distant from the area of the anastomosis but does not allow pooling. We also avoid the contact of celluloses or cotton gauzes directly with the lumen of the vessel due to their thrombogenic properties.

For the lumen irrigation, we use a heparin solution with 200–250 IU/ml [11]. Washing the lumen of the vessel directly can hydrodissect the vessel wall, exposing the subintimal collagen. Therefore, we introduce a blunt-tipped lacrimal cannula into the lumen of the vessel before anastomosis to perform a gentle wash [17]. Likewise, we usually do an irrigation of the flap through the artery with 20 or 30 ml of heparin solution, prior to the transfer; this checks the correct flow in the vascular circuit of the flap.

3.3 Instruments care

This delicate material requires little but precise care. First of all, we should avoid falls during surgery or washing, as the tips of the material can be damaged. If this happens, the closure of the material would not be perfect and its functionality would be noticeably reduced. It is also necessary to avoid the tips of the material to be oriented towards the sides in the store box, since movements with the box closed could also damage the tips inadvertently.

The material should preferably be washed by the scrub nurse or the surgeon himself, who is familiar with it and will be more careful with its handling. A final wash should be done with distilled water and dried with an air gun to prevent rust formation.

During surgery, the material must always be clean and moist, so that the sutures do not adhere to its surface. Dirty material and damaged tips will cause problems with the suture technique at key moments of the intervention.

3.4 Sutures

The most common sutures elected are the 9-0 on a 100-μm needle and the 10-0 on a 75-μm needle. Because of the ease of knotting and the low tissue reaction, the most used material in sutures is nylon. Some authors prefer polypropylene due to a lower tissue reaction, but its knots may be less reliable.

4. Microsurgery techniques

There is no stipulated standard on how to perform a microvascular anastomosis, the choice of the specific technique is operator-dependent. However, there are certain issues that we must avoid: a narrowing of the vascular lumen, an irregular distribution of the diameters of the vessels that would generate folds and irregularities, an excessive suture material inside the vascular lumen, and above anything else transmural sutures that bite the posterior wall closing the vascular lumen [24].

4.1 End-to-end anastomosis

By far, the most frequently employed technique is the end-to-end anastomosis. Because of its simplicity in less experienced hands, it has one of the lowest failure rates.

4.1.1 Triangulation

This technique was described by Alexis Carrel in the 1902. His intention was to separate the posterior wall from the anterior, as he realised about the danger of transmural stitches. The technique employed three initial sutures, with 120° separation between each [25]. It was modified with the use of only two initial sutures at 120° or 150° distance, as the posterior side was then longer and also fell away (**Figure 3**). Finally, it was modified again to propose only two initial sutures at 180°. The rest of the anastomosis will be closed with simple sutures between the initial points [24, 26].

4.1.2 Continuous

The continuous suture saves time and corrects discrepancies of 2–3 mm in size between vessels, but it has as an inconvenient: the tobacco bag effect. Some authors propose to distribute at first the two vascular lumens with some simple stitches. This technique is not very popular in venous microvascular anastomosis due to its stenosing tendency [24, 26].

4.1.3 Continuous interrupted

The continuous interrupted technique (also known as open-loop technique) is our technique of choice. It combines the safety of simple sutures with the comfort and speed of the continuous ones. It allows to constantly maintain a perfect visualisation of the vascular lumen and at the same time minimises the necessary manoeuvres. In this technique, a continuous suture with a spiral of very wide loops is made,

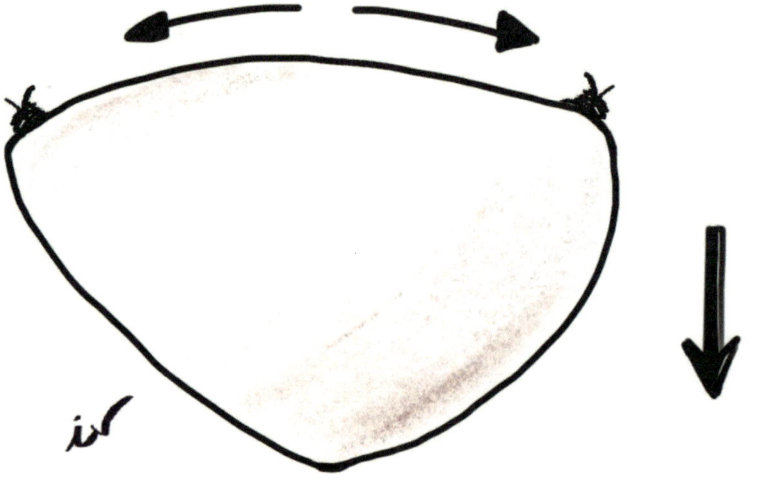

Figure 3.
Triangulation technique, after placing tension between the first two stitches, the longer posterior wall of the anastomosis falls down, precluding transmural stitches.

then moved towards a lateral. Finally, each one of the loops is sectioned and knit separately [24] (**Figure 4**).

4.1.4 One way up

This technique is of first choice when we cannot properly manipulate both the vessels of the microvascular anastomosis, we cannot manage to rotate it in order to carry out the suture of the posterior wall. When performing the one-way-up technique, we begin suturing the posterior side. The needle is introduced from the deep side of the vessel to the intima of the posterior wall and returns through the intima in the lumen of the posterior wall of the opposite vessel. The knots are the same as in simple stitches. After placing three or four stitches in the posterior wall in an inverted fashion, it is easy to perform the remaining stitches in a conventional way. It is important to place the posterior wall stitches close enough to prevent any leakages, as revising the posterior wall is bothering. Lastly, the anterior face is sutured. This technique is one of our preferences as it minimises the incidence of transfixing sutures [24] (**Figure 5**).

4.2 End-to-side anastomosis

This type of suture is very useful when there is a great discrepancy between vascular lumens, or when the flow through a vascular axis must be preserved.

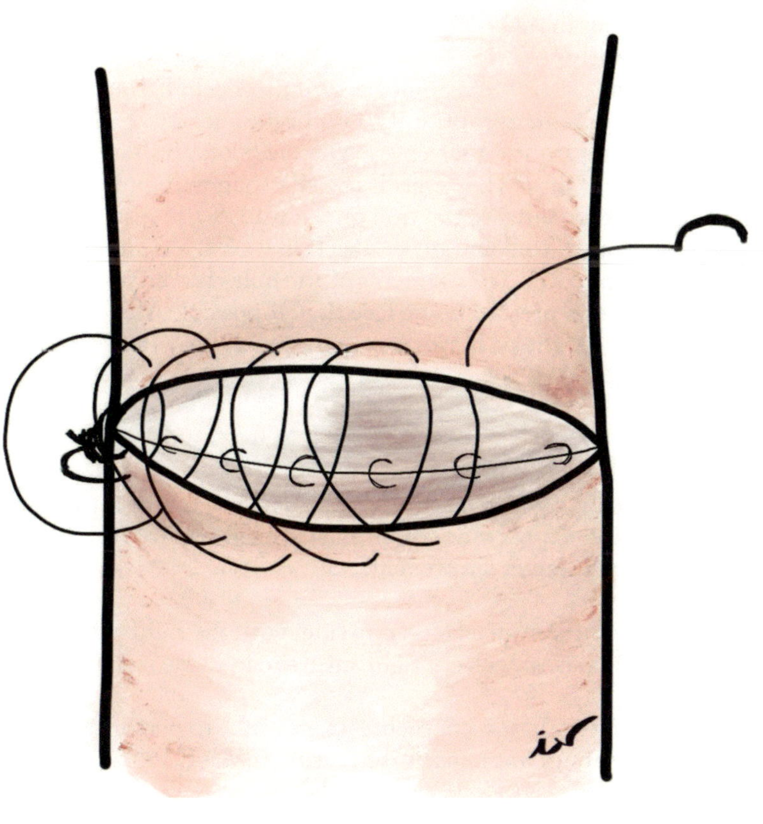

Figure 4.
Open-loop technique. Continuous suture of the upper face of the vessel, with very loose loops. Afterwards these loops are divided, and knotted as simple stitches. The posterior wall is depicted sutured first.

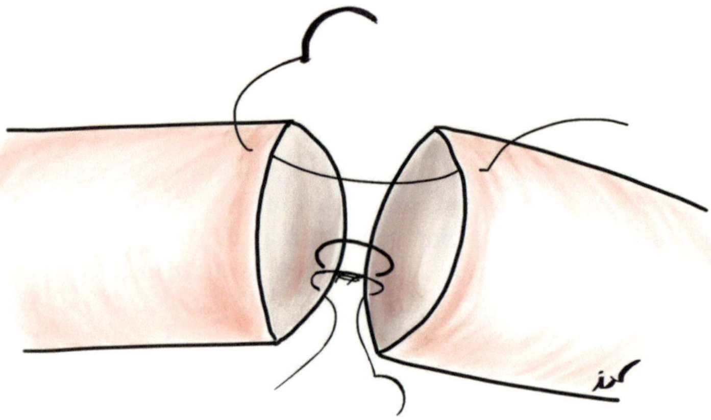

Figure 5.
One-way-up technique. First four to five stitches are placed in the posterior wall in an inverted fashion. It is important to leave only a small gap between the two first knots in this posterior wall, in order to avoid leakages and reviews here. After these first inverted stitches, the rest of them are placed in a conventional simple fashion as depicted in the figure. This technique avoids twisting and injuring the anastomosis.

Therefore, it is very useful in lower limb reconstructions, when one of the vascular axes is damaged or we want to preserve the integrity of all [27]. For example, in head and neck surgery, after a cervical dissection, the high rate of venous thrombosis makes it advisable to choose the internal jugular as recipient vein [6]. In view of the discrepancy between the internal jugular and the vein of any flap, as well as the pertinence of maintaining the flow through the internal jugular, an end-to-side anastomosis is frequently chosen.

To perform this end-to-side anastomosis, we must occlude the flow through the larger vessel that will remain in continuity. Our preference is the use of two rubber loops with a double pass around the vessel. When tensioning these loops, it seems that the damage to the walls of the vessel is inferior than with bulldog or baby Satinsky clamps. Next, by putting traction on the wall of the vessel with a transmural suture, we elongate the wall and make a section with the straight adventitectomy scissors or with a scalpel [27]. The diameter of the hole created must not be greater than the one on the vessel present in the free flap. If possible, the flap is tilted over the anastomosis to suture the posterior face; otherwise, we will use a one-way-up suture technique [28].

4.3 Tips and pearls

- It is important to take within each suture a good amount of intima to adequately evert it and expose smooth intima to the vascular lumen, with scarce subendothelial collagen or suture material.

- The knots should be flat, placed on one side, with the right pressure just to close the anastomosis, since very tight sutures can cause isquemia and failure.

- In case of working with veins of inconsistent walls, to perform an immersion technique, using abundant heparinised serum in the field to open the vascular lumen can be useful.

- We should not allow leaks in the anastomosis; these will cease through the formation of an intraluminal thrombus, which can ultimately endanger the entire anastomosis.

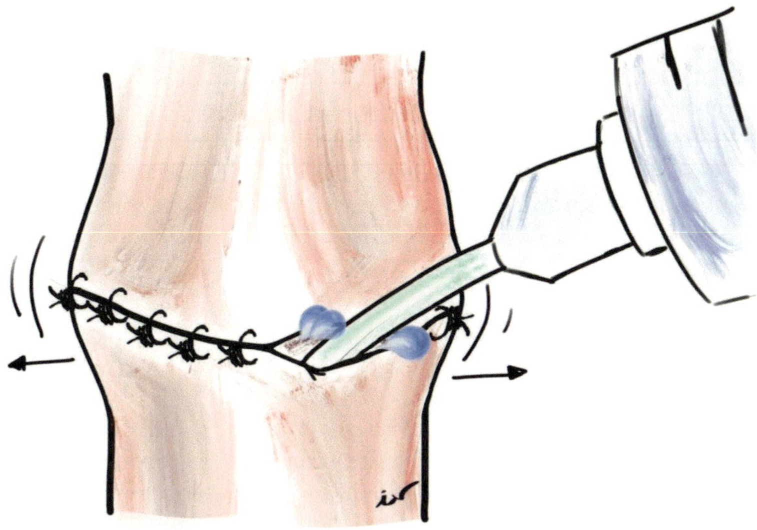

Figure 6.
Patency test with rinsing. A short Abbocath cannula is introduced in the microvascular anastomosis in between the space left by the two late stitches, we pretend to verify an easy dilatation before finishing the suture. The anastomosis is completed full of heparinised serum, until clamps are released.

- At the end of an anastomosis, we must check its permeability, for example by means of a patency test. Other possibility is to make a profuse irrigation through the space left in the microvascular anastomosis before placing the last two stitches, an inflation and slight dilatation of the anastomosis with the heparinised serum evinces the vascular patency (**Figure 6**). The classic patency test can traumatise the intima.

- The learning of these techniques must begin in a laboratory of experimental surgery with animal models [29].

- Dilatation of the lumen with specific dilator forceps allows better visualisation of the interior of the vessel, easily recovering the needle at each stitch.

- Before passing the entire thread through the anastomosis, the former should be in line with the vessels, and not angulated behind the needle. This precaution will avoid tears and friction on the vessel wall with the thread passage.

- Limit the vessel dissection as much as possible (**Figures 1** and **2**).

5. Couple devices

Since the onset of microsurgery, a great interest was drawn towards the development of suture techniques to perform anastomoses more quickly and automatically, in order to buffer inaccuracies [28]. For this purpose, devices in the form of two metal rings that are coupled, known as coupler devices, were developed.

Currently, its use is widespread, mainly for vein anastomoses, although they have also tested a 100% patency in arterial ones. The vessel is introduced through the ring and the edges are fixed inside-out in the pins arranged in the ring, then the same is done with the other vessel and the hinge of the device, that joins both sides, is closed. The eversion of the edges achieves less exposure of the vascular lumen to foreign material and therefore the rate of thrombogenesis is lower. This eversion of the edges in the case of the arteries is more complex due to the thickness of the vascular wall, which makes its use in arterial anastomosis not so popular [28, 30]. There are coupler devices currently available with built-in systems for flap control, such as Doppler.

Despite their many advantages, they present some drawbacks. Although they have been shown to reduce the time needed to perform the anastomosis, their use involves some complexity and produces some stenosis. On the other hand, they are not recommended in areas with a tendency to infection, with poor vascularisation or to be irradiated.

6. Conclusions

In the search of the best functional and aesthetic results, free tissue transfers have become the gold standard for many of the issues that arise in reconstructive surgery [31, 32]. Clear examples of this are the deep inferior epigastric artery perforator flap in breast reconstruction [7, 8] and the anterolateral thigh flap in head and neck reconstruction [6]. But the success of these techniques does not only depend on an adequate vascular suture, but also on a constellation of details that must be taken into account. These go from the availability of a trained team, to ergonomics, through a scrupulous cleanliness of the surgical field. All this does nothing but stress the importance of patience, good planning, attention to details or even the use of microsurgical checklists in to prevent any error that, however small, can have catastrophic consequences.

Acknowledgements

The authors would like to give special thanks to their chief Dr. Antonio Taboada-Suárez for his support, constant stimuli and generosity and to Dr. Joaquim Megías Barrera for his guidance.

Conflict of interest

The authors declare no conflict of interest.

Author details

Ignacio Vila*, Iván Couto-González and Beatriz Brea-García
Department of Plastic Surgery, University Hospital Complex of Santiago, Santiago de Compostela, Spain

*Address all correspondence to: ignacio.vila.garcia@sergas.es